SUPERCONTINENT

Ten Billion Years in the Life of Our Planet

TED NIELD

Harvard University Press
Cambridge, Massachusetts
2007

'Different living is not living in different places' by Stephen Spender from
New Collected Poems © 2004, reprinted by kind permission of the Estate
of Stephen Spender; 'Humming-Bird' by D. H. Lawrence, reproduced by
kind permission of Pollinger Limited and the Estate of Frieda Lawrence
Ravagli; 'On a Raised Beach' by Hugh Macdiarmid from *Complete Poems*,
reprinted by kind permission of Carcanet Press Limited; *The Hitchhikers
Guide to the Galaxy* © 1979 Douglas Adams, reproduced
by kind permission of Pan Macmillan UK.

First published in the United Kingdom by Granta Books, 2007

Printed in the United States of America

Library of Congress Cataloging-in-Publication Data

Nield, Ted.
Supercontinent : ten billion years in the life of our planet / Ted Nield.
p. cm.
Includes bibliographical references.
ISBN-13: 978-0-674-02659-9 (alk. paper)
ISBN-10: 0-674-02659-4 (alk. paper)
1. Pangaea (Geology) 2. Continental drift—Computer
simulation. 3. Lost continents. I. Title.
QE511.5.N54 2007
551.41—dc22
2007008205

The mind must believe in the existence of a law, and yet have a mystery to move about in.

JAMES CLERK MAXWELL

CONTENTS

ACKNOWLEDGEMENTS

For particular help with this book and with previous writings of mine that have contributed to it, I gladly acknowledge the following persons (who are, of course, in no way responsible for remaining omissions and errors, for all of which responsibility rests with me).

Professor Philip Allen, Professor Mike Benton, Ms Vivianne Berg-Madsen, Professor Kevin Burke, Dr Tony Cooper, Professor John C. W. Cope, Professor Charles Curtis, Professor Ian Dalziel, Dr Wolfgang Eder, Professor Michael Ellis, Professor John Grotzinger, Dr Gordon Herries-Davies, Professor Paul Hoffman, Mr Robert Howells, Dr Patrick Wyse Jackson, Dr Werner Janoschek, Dr Sven Laufeld, Dr Roy Livermore, Dr Bryan Lovell, Dr Joe McCall, Professor Mark McMenamin, Dr John Milsom, Professor Eldridge Moores, Dr Bettina Reichenbacher, Professor John J. W. Rogers, Dr Mike Romano, Professor Mike Russell, Dr Gaby Schneider, Professor Chris Scotese, Professor Dick Selley, Professor Bruce Sellwood, Professor Dr Klaus Weber, Dr Jeffrey Huw Williams, Mr Simon Winchester and Dr Rachel Wood. My special thanks go to those in this list who critically read parts of the book in manuscript.

I should like to acknowledge the Geological Society of London for its enlightenment in encouraging private enterprise among its employees; but I also owe an immense debt to the Society as a Fellow. Fellowship has provided me with invaluable access to one of the great geological libraries of the world; and to the services of my colleague, Wendy Cawthorne. Wendy, like all the best Assistant Librarians, assists in finding the things readers ask for, but then goes the extra mile to find the things they actually need.

The idea for this book came to me very early one happy summer morning in 2003, among the chestnut trees of Vallée Française, Lozère, France. I made the first outline in a letter I was writing to my dear friend since student days, Professor Mike Ellis, now at the US National Science Foundation. Had he and I not been corresponding in this old-fashioned way since he selfishly removed himself to the other side of the Atlantic, I might never have begun this project. I also thank my editor, George Miller at Granta, who took me to lunch and made editorial suggestions that greatly improved the text.

I should pay homage to the late and great Professors Janet Watson and Mike Coward of Imperial College, London. They never taught me in the strict sense, but after reading their work as a student I eventually met Janet and came to count Mike as a friend. In this group must also be numbered Dr Rod Graham (still vigorously extant), who did teach me, but who has since, I hope, forgiven me. To all I owe my sense of awe at their achievements in untangling the rocks of the Precambrian. I must also acknowledge two more of my personal giants, the late Professors Derek Ager and Dick Owen, both of whom taught me by example that the most complicated science ought to be explicable in language everyone can understand: a lesson that stood me well in my subsequent career as a science journalist.

I hope that this book will be seen as one long homage to all those great geologists whom I have met over the years and who have helped me. I lay no claim to having seen further, but in the thirty years that have passed since I began studying Earth science, too many giants have offered me their shoulders as footstools for me to be able to acknowledge them all by name. However, for the dedication of this book I would like to single out my fellow members of the Management Team of the International Year of Planet Earth, with whom discussions on the way that Earth sciences benefit society in general have played a major role in the development of this book.

Most of all, my thanks go to my wife Fabienne, who has continued to provide that without which nothing would be possible.

Ted Nield

FOREWORD

BIG CRUNCH

Different living is not living in different places
But making in the mind a map.

STEPHEN SPENDER

The drifting continents of the Earth are heading for collision. Two hundred and fifty million years from now, all landmasses will come together in a single, gigantic supercontinent. It already has a name (in fact, it has three) even though human eyes will, in all probability, never see it.

That future supercontinent will not be the first to have formed on Earth, nor will it be the last. The continents we know today – Africa, the Americas, Asia, Australia, Europe and the Antarctic – are fragments of the previous supercontinent Pangaea, which gave birth to dinosaurs, and whose break-up was first understood barely a century ago, in 1912. Yet 750 million years before Pangaea formed, yet another one broke up; and before that another, and so on and on, back into the almost indecipherable past. The Earth's landmasses are locked in a stately quadrille that geologists call the Supercontinent Cycle, the grandest of all the patterns in nature.

Men and women have been imagining lost or undiscovered continents for centuries. For early mapmakers they filled in gaps, forming a bridge from the uncertain to the unknown. Nineteenth-century zoologists and botanists speculated about sunken lands to explain odd distributions of animals and plants. Early evolutionists peopled their hypothetical lost lands with the ancestors of mankind. Fringe religions adopted them and embattled minority cultures latched on to them to bolster their myths. All had one thing in common: the basic human urge to understand and make sense of the world.

Today geography has no room for lost continents. The world is ringed by satellites that reveal no undiscovered country. But lost continents have found, at last, a true science of their own. This book is about how that science emerged and how Earth scientists are using the most modern techniques to wring as much information as they can out of the oldest rocks on Earth and predict what the next supercontinent will look like.

Supercontinent Earths, salvaged from oblivion or projected into the future by today's geologists, share one thing with all the lost continents that were ever dreamt of, whether by other scientists, mystics or madmen. All lost lands truly exist only in one place: the human mind, the only eye that can see through time.

But why should we care? We human latecomers evolved a mere six million years ago, halfway through the present cycle, when today's moving continents were barely a few hundred kilometres from where they are now. And if what we understand of other species can be applied to ours, there is very little chance that humans will survive the 250 million years that will pass before a new supercontinent assembles.

Yet the supercontinents of modern geology are no exotic fruit from some esoteric branch of science. Their discovery began with an innate urge to explore; it was boosted by the spur of Empire in the nineteenth century as science reached out through the third dimension to map the world and its living things. It continued as the patterns of today were seen to hold meaning for their evolution through the fourth dimension, time. And as the human mind has reached out it has also drawn together.

Without science the Earth could not sustain us in anything like our present numbers. Our continued life on the planet that gave rise to us will depend upon our ability to use our science to protect and feed ourselves in the face of what threatens us (chiefly ourselves). Understanding the Supercontinent Cycle is nothing less than finally knowing how our planet works. This can be to our benefit – we have, after all, made it thus far – or our detriment.

If scientific knowledge had been properly deployed many, perhaps most, of the quarter of a million people who died around the Indian Ocean on Boxing Day 2004 could have been saved. The knowledge that makes that possible is the same knowledge that reconstructs landscapes that washed into oblivion hundreds of millions of years before our species existed.

London, 2006

PART ONE

MOVING IN MYSTERY

1

LOST WORLDS

Far out in the uncharted backwaters of the unfashionable end of the Western Spiral Arm of the Galaxy lies a small unregarded yellow sun. Orbiting this at a distance of roughly ninety-two million miles is an utterly insignificant little blue green planet . . .

DOUGLAS ADAMS

Novopangaea – a science fiction

A blue planet hangs in space. You have seen many planets as you have searched the cosmos for signs of life far from your own small planet somewhere in the vicinity of Betelgeuse. But as you approach this one, something about it impresses and excites you. It's the third planet from an unremarkable star, and the largest of the rocky inner ones. But as you approach it from below the plane of the ecliptic, it shines like an opal, streaked with white.

The galaxy is full of the common oxide of dihydrogen that appears to cover this planet, but almost everywhere else it is a solid. Here it exists as a liquid and there are traces of its vapour in the atmosphere. The liquid phase can only exist within a very small range of temperatures; temperatures you, as a space explorer, expend a lot of energy maintaining inside your craft. Yet here these equable conditions seem to cover the entire planet. There isn't even an icecap at the pole, where the temperatures should be at their lowest. It is almost inconceivable that a planet's temperature should be so constant over its entire surface. It must be that the deep atmosphere is trapping the star's heat, and then, with the help of the ocean, spreading it around.

Above the glowing blue ocean, especially over its equator, are streaks of white. Cloudy curlicues and spiralling weather systems track like miniature galaxies across the hemispheres. At first it all looks chaotic, but on your long approach, heading towards the planet's southern pole, you have time to study time-lapse images. Suddenly the apparent chaos starts to make sense. The clouds' movements are indeed complex, but do describe a sort of ragged mirror symmetry about the planet's equator. What seemed like chaos now looks more like order: the atmosphere is convecting in six great cells arranged symmetrically about the equator.

The planet's moon is unusually large, though to an experienced space traveller little else is unusual about this satellite. No heat regulation there; with no atmosphere to distribute energy, temperatures can swing wildly through almost 300 degrees from sunlight to shade: quite normal for a space rock struck by starlight. A satellite as big as that must set up a tidal bulge in the ocean by the force of its gravity; you will be able to detect that once you are in orbit and can train your altimeter on the ocean surface.

Already, using the spectrometer to analyse the light reflecting from the planet, you have detected, amid the dominant nitrogen signature gases like carbon dioxide, and the gas phase of dihydrogen oxide (which will also help to trap heat and keep the planet warm). Methane is there too, and does the same job.

The unusually tall oxygen spike piques your interest, but just as you are thinking about that, something momentous distracts you. Your ship is now pulling level with the planet's equator. Perhaps because the clouds had drawn all your attention you had missed it at first, but now you see that this is not a liquid-covered world after all. There, below, is a single, gigantic landmass. You can see it clearly, because the clouds obligingly part over it; few penetrate far beyond its coastline. As the hours go by you watch the landmass unroll as you enter a fixed equatorial orbit.

It sits mostly in the Northern Hemisphere, covering perhaps 30 per cent of the planet's total surface area. It is dry. Immense white and beige deserts occupy nearly all of it. Three ranges of mountains, low, desolate and worn down with age, stand out amid the dune seas and endless dazzling playas. Dry, wiggling canyons feed the arid interior wastes, dying into vast plains of white from which expanses of blown sand stretch far away beyond the shimmering mirages of the horizon.

Most spectacular, apart from this terrifying barren waste, is the continent's southern coast, maybe twelve or fifteen thousand kilometres long. It lies at a slight angle to the equator and crosses it near its southern end before taking a dogleg and heading back, reaching even greater heights, to the north-west. This entire coast presents a cordillera of jagged peaks up to eight thousand metres high (perhaps nearer ten thousand at its eastern end) punching into the cold upper atmosphere and capped with white. These mountains are young, active, still growing. There are volcanoes too. One of them is erupting now, its plume of ash sweeping offshore like a thin veil, carried by the winds of the topmost atmosphere. This planet's surface is moving, geologically active; the planet is alive inside, powered by heat generated continuously by radioactive elements, so that the whole crust seethes like the scum on a boiling pot. As the largest of the rocky planets in this system, it is big enough not to have cooled down and died like the others, even after nearly five thousand million years.

What excites you perhaps more than anything as a space explorer are the colours you can see at the coast of the supercontinent, especially where those coasts cross the tropics. The interior is parched; but the point at which the driest area comes closest to the ocean is behind the range of towering mountains on the south-east coast. Here the weather systems that sweep inshore stand little chance of breaching those massive battlements (even though some of those systems are thousands of kilometres across, with wind speeds of over three hundred kilometres per hour).

But on the diametrically opposite north-west coast things are different. Here, where prevailing westerlies make landfall, streamers of cloud obscure the land for thousands of kilometres. Beneath them, from time to time and around the edges of the cloud blanket, you detect a livid green stain. Other, narrower areas on the supercontinent's coasts are green too, peeping out occasionally from the fringe of coastal clouds. This is the eureka moment. The oxygen spike! That huge ocean, and those coastal regions where moisture falls as rain, are teeming with living things.

You have just topped the greatest scientific discovery by any member of your species since it first began to look up towards the stars. You have found another place in the universe where matter *lives*. You knew it must be possible. Some said probable. Growing numbers believed it inevitable. But would – indeed, *could* – anyone ever *find* such a place? Given the distances of space, would it be possible to travel to such a world? And then even if, somewhere else in that limitless abyss, matter had become imbued with life, would it necessarily *coincide* with yours? For there was another abyss to consider: the abyss of time.

The universe is like a post-apocalyptic town: there appear to be other houses, but only yours is currently inhabited. Maybe all those other living worlds were marooned not only by impossible and untravellable distances, but also by *duration*; lost in time as well as space. Now you have an answer. You have found a neighbour alive.

You learn as much as you can about this precious place from orbit, but the next thing on your mission directive is to check if any among those living things down there is sentient. But you already know the answer to this question. Sentient life becomes civilization in a geological instant, and the chances of finding the first living planet at just that tiny moment between the evolution of an intelligent being and its ability to build cities and get power from atoms are too small to imagine. There are no satellites orbiting. There are no transmissions. This planet hangs in space like a great unseeing eye. There is no civilization down there. The creatures that may swim in its seas, or fly through its air, wander those forests or cling to its fertile coasts are dreaming their innocent world, unaware of anything beyond it, or that over them all, your shadow has now fallen. It is a kind of paradise. You envy it.

But hunches are not enough. Rules are rules and the manual says you have to check, run tests, write reports. From your vantage point, with your instruments, you can now scan the surface of this planet's landmass in precise detail, centimetre by centimetre. If there is (or was) a civilization down there, you will find it. Even if some extinct creature had built something, or carved the sacred images of its great leaders into some granite mountain, you will see it.

You begin the scans, which eventually will be assembled in a computer that will remove all the obscuring clouds; but you know this is hopeless. If the absence of transmissions tells you there's nothing intelligent there now, the planet's reflected light tells you there was probably *never* anything there. Sentient life quickly learns the secrets of matter and makes power from atoms. That process creates forms of matter that never exist in nature and which take millennia to decay away to nothing. Even on an active planet like this one, with weathering and erosion and deposition and a seething crust that renews and recycles itself, these substances endure. They are the only truly lasting products of civilization.

You scan the arid surface of the supercontinent for radioactive isotopes of the most insoluble elements with the longest half-lives. You find some: Thorium 242, and Uranium 235 and 238. But these all occur naturally. There is nothing more. If there ever had been an advanced civilization on this planet it must have vanished more than 100,000 years ago, though this does not yet depress your archaeologist, because the surface scans are not finished.

Much of the land surface is dry. Physical traces of civilization might have survived for more than 100,000 years, a great city perhaps, or some massive monument hewn from living rock, whose outline would still be visible. But even after you have assembled in your memory banks a complete inventory of every valley, mountain and hill on this supercontinent, the archaeologist finally admits defeat. Now only the geologist is interested in the possibility. But looking for fossils is something you cannot do from a spaceship.

As the weeks go by, you turn (with little real enthusiasm but because the mission directive says you must) to the planet's moon. You have seen so many other bodies like it, a dull, cratered space rock, dry, dead, circling for ever, its inert surface open to space, almost unchanging from its earliest violent days of heavy meteorite bombardment.

Yet here you are in for a surprise. Almost immediately your preliminary scans of the surface turn up real and unequivocal evidence of advanced civilization. At six separate sites you identify the remains of landing craft, transportation vehicles, transmitters, a seismic array. Another enormously significant revelation for your mission. Not only is your home world no longer the only living planet in the universe; you know now there have been other space explorers; and what is more, they have passed this way.

Subsequent archaeological research on the lunar artefacts reveals the trail to be a little cold, however. As the moon is clearly not itself a living planet, you can check out one of the alien spacecraft. Landing some way off and approaching with care, you see it and, most exciting of all, the footprints around it. They still look as though they arrived yesterday: unequivocal evidence of alien beings with spaceflight capability. Your archaeologists take samples of the spider-like landers, from each of which an escape module once blasted into space. Clearly the six visitations to this moon were made by the same beings, at more or less the same time. But when?

The answer is not long in coming; microscopic examination of the metal surfaces reveals that they are pitted with billions of tiny micro-meteorite impacts. From their density, and what you know about the rate of influx of such tiny objects, you work out that these visitors – whoever they were – must have left for the last time 200 million years ago. But there's a nagging question. Why would these travellers come to this moon, and not to the much more interesting planet?

Perhaps, despite the lack of any trace of them there now, these beings had come *from* the planet?

You ask your geologist, who has been studying the planet's single continent and its volcanic mountain chain. From what she now knows of the planet's crust and how it moves, even if some civilization of 200 million years ago had completely covered that same planet in cities and then wiped itself out in some gigantic global nuclear holocaust, nothing – not even the faintest trace of some unnatural radioisotope – would now remain on the surface. What is more, if those vanished beings were to be brought back today, they wouldn't recognize the world below as theirs. At the speeds at which the planet's crustal plates move, even with all the land locked together in a great super-continent, she can be certain that 200 million years ago the planet looked nothing like this. Perhaps then there were once many smaller, separate continents, all scattered about like islands in the ocean.

The geologists begin to write a research proposal to break the mission directive and visit a living world for the first time, spurred on now by the possibility of finding fossils of a vanished sentient race that may have developed space flight before vanishing completely. Now the only trace of them and their culture could be six short visits they had made in their heyday to their dead, unchanging moon, lasting in all not much more than 300 hours.

And who knows what they would find if they got permission? Maybe those alien explorers are in for yet another shock. Perhaps those fossils that they discover of a small, forked creature would look very familiar – just as the footprints on the moon had done. Perhaps the space visitors from a small planet in the vicinity of Betelgeuse would find themselves meeting their ancestors. Perhaps they would discover themselves to be one lost tribe within a galactic diaspora that had saved the human species from inevitable extinction on a home world to which it had now, for the first time, returned.

Future worlds

Scientists are already trying to predict what this supercontinent of the distant future will look like, and the version I have just described is based on the work of a British scientist who divides his time between the chilly waters of the South Atlantic and the British Antarctic Survey's headquarters in Cambridge. Roy Livermore is a marine geophysicist. He is interested in computer-modelling the way the plates of the Earth's crust move, and his main research area is the stretch of ocean floor between the tip of South America and the Antarctic Peninsula. As Roy pointed out to me these two eastward-sweeping points of land, Graham Land and Tierra del Fuego, they reminded me of a section through a piece of armour plating pierced by a high-velocity round. The hole through which the bullet appears to have passed is known as Drake Passage.

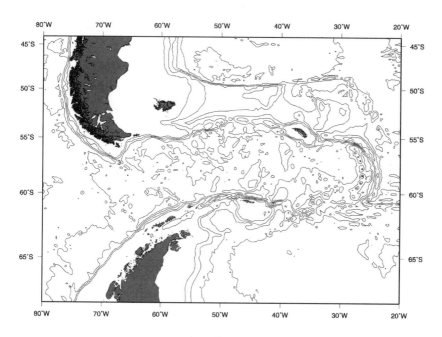

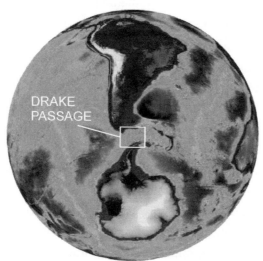

Drake Passage, between the southern tip of South America
and the Antarctic Peninsula.

Knowledge about how the Earth's tectonic plates have moved since Drake Passage opened up thirty million years ago has many uses. Oil companies are interested in how continents break up and move because fertile environments for oil formation are created at the continental margins. Climate modellers are interested in the Drake Passage because its opening contributed to a climate switch some thirty million years ago from the warm 'greenhouse' Earth to the cool 'icehouse' world we live in today. But Roy's theoretical interest in computer-modelling plate motions has recently enjoyed a more unusual application. For Roy Livermore created the future world of Novopangaea (the name he chose for it reflects the return of a condition that prevailed 250 million years ago), when all the present continents were last assembled into a single landmass, known to geologists as 'Pangaea' ('All land').

His vision of the deep future was the setting for a TV documentary series about the likely course of animal and plant evolution. Its producer, John Adams, wanted to show what the Earth might be like five, 100 and 200 million years from now, and particularly how the animals and plants of that world might look. The focus of the series was principally the animated living forms; but on what kind of a world would these CGI creations disport themselves?

Adams began by asking geologists what the world would look like in the future. Livermore had to create a set of credible plate-tectonic forward projections or 'preconstructions' of our future Earth. In other words, he had to put the continents into the positions they would occupy in five, 100 and 200 million years from now. Climate experts then took his maps and predicted how the atmosphere and oceans would behave given those arrangements of land and sea, and so deduce what conditions could be expected. The resulting future world could then be populated with appropriate fauna (including 'flish', flying fish, and a rather attractive hopping snail) dreamt up by evolutionary biologists.

'It really grew out of work I did with Professor Alan Smith in the Department of Earth Sciences here in Cambridge, back in the eighties. I have been interested in preconstruction and thinking about the future for maybe fifteen years,' he told me, producing three Lambert projections of the globe, the land in green, ocean in light blue and shelf seas in a darker shade.

So how did he arrive at Novopangaea? Is it simply a matter of knowing how the continents are moving now and winding the clock forward on a computer? 'Well, first of all this exercise isn't driven simply by scientific curiosity,' he explains. 'There were other considerations like, from my point of view, the desire to illustrate a range of geological processes.' In other words, in making his preconstruction, scientific constraints were mingled with the need to arrive at an interesting outcome for the programme makers. This means there has to be a point when forward projection ceases to be mechanistic and objective; when the experimenter must intervene. (As we shall see later, the imaginative ambitions of *all* those who have ever dreamt about supercontinents, past or future, have rarely been unmixed.)

Livermore took me through the process. 'We start with the present day, when we know how the plates are moving, and extrapolate a few million years into the future. The five-million-year projection is quite tightly constrained. It's what you get if you just wind everything forward a bit.' Because of that, the outcome is not terribly exciting; nothing seems to have changed very much. 'No, nothing terribly exciting happens until we get to 100 million years. But whenever you go that far, there inevitably comes a point where you have to make a decision.' This is where the model operator plays God. The planet is not a simple perpetual-motion machine that cycles for ever in the same old way. From time to time unique events happen that alter the outcome of the process, like a massive meteorite strike, or super-volcanic eruption. These events are often unpredictable, but the Earth carries their consequences for all time. Understanding this has been a major breakthrough in our thinking about the Earth in the past 200 years.

'The biggest "decision" I made,' Livermore told me, 'was that the Atlantic Ocean would continue to open and the Pacific would continue to shrink.'

All land

The continents of today's Earth are the wreckage of that single supercontinent, Pangaea, which began to break up about 250 million years ago. The name was given to the last supercontinent to have formed on Earth by German geophysicist Alfred Lothar Wegener. It first occurs in the third edition of his great book *The Origin of Continents and Oceans*. When it first appeared, in 1915, it was the first serious attempt by a modern Earth scientist to convince the world that continents drift.

As we shall see, for any imaginary supercontinent to catch the imagination of scientists and public, its most important asset is its name. However brilliant, instinctive and insightful Wegener was as a geophysicist, he understood and cared little for public relations. This was a shame, because had he understood and cared about PR a little more, his theory might well have fared a lot better than it did, especially in America.

Today, when Pangaea ranks as high on the romance scale as such exotic names as Ushuaia or Zanzibar, it is amazing that Wegener should have introduced it so casually in the final chapter of his third edition. While writing about the single landmass he believed brought all today's continents together, he simply says, 'This Pangaea . . .' inserting it as a synonym for the long-winded explanation in the sentence before – a synonym he clearly expects his readers to have enough Greek to understand.

The third edition of Wegener's book was the first to find an English publisher (Methuen & Co. of London). They engaged one John George Anthony Skerl as its translator, and he changed the Germanic spelling Pangäa to Pangæa, so must therefore be credited with bringing the word into the English language, in 1924. English palates soon found it easier to pronounce it *'panjeea'*, because the same Greek root – *Ge*, meaning 'earth' or 'land' – also gives us 'geology'. Finally the Americans, dispensing with the archaic ligature *æ*, decided to spell it 'Pangea'.

Clinching evidence that Wegener cared next to nothing about his new term is provided in the fourth (and last) edition of his *magnum opus*, published in 1929. This edition had to wait thirty-seven years before being translated into English and published in the USA. In this edition, however, there is no 'Pangaea' in the index; nor in the speculative chapter on the forces that might cause continents to drift; nor anywhere. Wegener just left it out.

Pangaea consisted of two smaller supercontinents joined at the hip in the region of the Equator: Laurasia in the Northern Hemisphere (North America, Greenland, Europe and much of what is now Asia) and Gondwanaland in the Southern Hemisphere, comprising South America, Africa, India, Australia and Antarctica. The world we see today is no more than Pangaea's smashed remains, the fragments of the dinner plate that dropped on the floor.

The main event in this slowest of all unfolding dramas was the opening of the Atlantic Ocean, which split North America from Asia and South America from Africa; though there were many other splits too. India emerged like a slice of pie from where it had lain wedged between Africa, Antarctica and Australia for the best part of 500 million years, drifted northwards across the Southern Ocean and smashed into Asia to make the Himalayas. Australia rifted from Antarctica, taking the Great Australian Bight out of its southern coast, and headed off for South-East Asia. Africa moved north and collided with Europe, a continuing process that will one day close up the Mediterranean.

These created a whole set of young oceans whose floors are nowhere older than about 250 million years: the date when Pangaea's rifting began. These ocean floors are forming along spreading centres like the Mid-Atlantic Ridge, scars that mark the original junction between rifted continental fragments, either side of which the ocean floor spreads away, carrying the increasingly distant landmasses with it at about the same speed your fingernails grow. Because the Earth is not getting bigger, one ocean expanding means another is shrinking. This is why, all around the Pacific, the ocean floor is being sucked back down into the planet in a process called subduction.

That, in essence, is plate tectonics. Because we now know the age of nearly every bit of ocean floor all over the world, it is relatively easy to see how the split took place. The ocean floor is, in effect, a road map showing how the continents have moved into their present positions, like the concentric ridges on the growing plates of a turtle shell.

If you feed all this information on continental trajectory and speeds of motion and rotation into a computer program like Atlas, the package Roy Livermore helped develop with Alan Smith and others at Cambridge University, you can animate the whole process and watch it unfold before your eyes. It is then relatively easy to run the program forward a little. That is why Livermore can be fairly certain about the way the Earth will look in five million years, the near future to a geologist. It is as objective as a computer model can be. The Atlantic will be a little wider, and Africa will be closer to Europe and Japan to North America.

But in order to look further and see how the drifting continents, riding the backs of convection currents flowing in the hot rocks of the Earth's mantle beneath them, will one day recombine, this is not enough. Sooner or later even plate-tectonic motions, so seemingly inexorable, must take a step change. The geologist has to intervene in the model with some educated guesses about these crucial turning points.

Roy's decision that the Americas will go on heading west, eating up the Pacific as they go, and crash into the amalgam of Asia and Australia, follows one interpretation of how supercontinents can form. Because it involves all the fragments of a previous supercontinent flying away from one another until they meet again on the other side of the globe (turning the previous supercontinent inside out), this process is called extroversion.

On the other hand, another process might apply. The old scars of the Earth's continents are lines of weakness, and for that reason history along them tends to repeat itself. The Atlantic is not the first ocean to have opened between the old continental kernels at the heart of Europe and North America. Several hundred million years ago North America was separated from northern Europe by a wide ocean, much as it is now. This seaway eventually closed to form mountains that were probably as tall as the Himalayas of today. On this side of the present Atlantic we see their eroded remains in Wales, Scotland and Scandinavia. Take away the present Atlantic and this old chain marries up to another old mountain chain in the eastern US. It was split in two when the Atlantic opened through it, roughly (not perfectly) along the same line. It was such geological evidence as this that helped the early proponents of continental drift demonstrate that their impossible-sounding theory might have some truth in it.

In other words, oceans can open and close, like a carpenter's vice, more than once. Imagine that you open a vice, put the carpenter's lunch (cold lasagne) into it and squeeze it tight. The lunch will ooze out and up, forming a mountain chain, which we shall call the Lasagnides. You then leave it until the lasagne has gone hard before opening the vice again. By now agents of erosion – mice – have scoured the once mighty Lasagnides back to bench level; but their roots, within the vice itself, remain. If you now reopen the vice to start the process again, some of those old Lasagnide remnants will stick to one jaw and some to the other; but the vice reopens along the same basic line. That is how you get some parts of the same mountain chain in Europe and others in America.

This is the second way to form a supercontinent; one that splits the landmass only to replace the pieces roughly where they were before. Supercontinent theorists call this introversion. Any would-be modeller of the next supercontinent faces this crucial question: will the Atlantic go on expanding and become a new 'world ocean' through extroversion, or will it eventually, like some tail-eating Leviathan, destroy itself by introversion? Roy sees no reason why the Atlantic should not continue to open; though other 'preconstructions' beg to differ, and for them the next supercontinent looks quite different from Novopangaea.

One preconstruction of the next supercontinent, created by Professor Chris Scotese of the University of Texas at Arlington, assumes that the Atlantic will one day close again. I put this idea to Roy Livermore. 'What Scotese is saying is that subduction is starting up in the Caribbean and the trenches of the Scotia Sea, and that these will propagate.' It is true; although the western half of the Atlantic Ocean's floor is welded tight to the eastern seaboards of North and South America, there are some places (in the Caribbean and between South America and Antarctica) where there is subduction. Livermore's research centres on the Scotia Sea area, however, and he doesn't see subduction there propagating up the eastern coast of South America. 'There is rapid subduction in these places, but it is tending to propagate eastwards,' he says. In fact, that eastward propagation is what makes Drake Passage resemble the pierced armour plating. What has passed between the two headlands is a narrow slice of ocean floor.

'Plus, it's still a moot point in geophysics as to how you start subduction off. How you do this at a passive margin, where the ocean crust is welded to the continent edge, is really not clear. Around the Pacific you already have well-established subduction zones that have been going on since the Permian. Why would they turn off?'

Livermore shifts his attention to the preconstruction he made for 100 million years hence, halfway to Novopangaea. 'To show how continents can rift,' he says, 'I have taken the liberty of opening up a new rift in here . . .' and his pen follows a new seaway connecting the Indian Ocean with the North Atlantic. 'We know the East African Rift is active, so we propagate that into the future by opening a small ocean. East Africa and Madagascar have moved across the Indian Ocean to collide with Asia; Australia has already collided with South-East Asia.' South of what is now India a mountain chain has arisen along a new subduction zone. And just south of it lies a familiar landmass, in an unfamiliar position. It is Antarctica. 'I don't believe Antarctica is going to stay at the pole,' he says. 'I *want* it to come north. Every other fragment of Gondwana has done that, piece by piece, and in the future Antarctica will; but only if it's dragged north by a subduction zone.'

Meanwhile the Pacific continues to shrink. North America and South America begin to wrap around the coasts of Asia. Australia has already collided with Japan and stuck. North America collides first, and South America sweeps around to consume the last vestiges of the Pacific and finally form Novopangaea.

Tales of Hoffman

Livermore is only the latest of a number of distinguished geologists to speculate about how today's continents might eventually recombine. The first to do so was Paul Hoffman. He called his future supercontinent, an amalgam of America and Asia, Amasia.

Much has been written about Paul Hoffman, and such accounts usually begin by remarking that he was the first on the scene. Writer and palaeontologist Richard Fortey has written: 'If expertise is defined as knowing more and more about less and less, I am at a loss to describe what it is to know more and more about more and more, but that is the Hoffman condition.'

One of the leading geologists of our age, Paul Hoffman is a man of whom stories are told. With his tanned, ascetic head, flashing eyes, mane of white hair and flailing, wiry arms, he has never been known to take prisoners. One geologist who has worked with him put it to me succinctly when he said, with a smile and a shake of the head: 'Paul is one hell of a scary dude.'

Hoffman is now Sturgis Hooper Professor of Earth Sciences at Harvard University. Much of his career, however, was spent at the Geological Survey of Canada. As one of the foremost thinkers on how plate-tectonic processes form supercontinents, and on how it might be possible to reconstruct supercontinents before Pangaea, Hoffman was also the first to write about the supercontinents of the future.

He, like Roy Livermore, thought that the Atlantic could well continue to expand and that we might be now in the middle of a process of continental extroversion (turning Pangaea inside out). He presented his idea at the 1992 Spring Meeting of the Geological Society of America in Montreal. 'The Americas are swinging clockwise about a pivot in NE Siberia,' he wrote in his abstract. 'They seem destined to fuse with the eastern margin of a coalesced Africa+Eurasia+Australasia, instituting the future supercontinent "Amasia".'

Hoffman used this pioneering preconstruction as a means of explaining something he believed about how supercontinents had formed and broken up in the deep past: the process of extroversion, turning old supercontinents inside out. He never published a map of Amasia, though in essence it might have looked something like Livermore's Novopangaea. However, he did give it a name, and a good one.

Amasia is often referred to by Earth scientists as shorthand for an extroverted resolution to the current pattern of Pangaea's break-up; but the concept received relatively little media coverage and so never really escaped into the wider world. Not so, however, Chris Scotese's projection, based on the opposite assumption, that the Atlantic will one day close back on itself. This creature definitely got out. And like all supercontinents whose names run amok, it has attracted the attentions of some strange and mystical colonizers.

Pangea Ultima's creator, Professor Chris Scotese, is a bear of a man with a big beard and a big smile. He has been involved for much of his career in reconstructing the continental positions in the past – a subject called palaeogeographic reconstruction – and in the Paleomap Project, an amiable, eccentric, homespun (and award-winning) website, www.scotese.com.

Scotese has produced a series of palaeogeographic atlases since he was an undergraduate at the University of Illinois in Chicago. His first were published as miniature 'flip books' in the 1970s and computer animations in the 1970s and 1980s. While a graduate student in the University of Chicago and the Paleomagnetic Laboratory at the University of Michigan, he and his supervisors published a series of maps that uniquely combined plate tectonics, palaeomagnetism and palaeogeography. These early publications of the Paleogeographic Atlas Project were the first to illustrate, through the emerging understanding of plate tectonics, how ocean basins and continents have evolved over the past 542 million years of Earth history.

Scotese's work did not hit the media, however, until 2000, when the NASA publicity machine published an interview with him about his work on the Paleomap Project. Attention focused on his ideas about the next supercontinent.

'We don't really know the future, obviously,' he told NASA science writer Patrick Barry at the time.

All we can do is make predictions of how plate motions will continue, what new things might happen, and where it will all end up. The difficult part is the uncertainty in new behaviours. If you're travelling on the highway, you can predict where you're going to be in an hour; but if there's an accident or you have to exit, you're going to change direction. And we have to try to understand what causes those changes. That's where we have to make some guesses about the far future 150 to 250 million years from now.

Among those predictions: Africa is likely to continue its northern migration, pinching the Mediterranean closed and driving up a Himalayan-scale mountain range in southern Europe. Of that everyone seems certain. Australia is also likely to merge with the Eurasian continent. 'Australia is moving north, and is already colliding with the southern islands of South-East Asia. If we project that motion, the left shoulder of Australia gets caught, and then Australia rotates and collides against Borneo and south China – much as India did 50 million years ago – and gets added to Asia.'

So far Scotese's vision works out very similar to Livermore's and Hoffman's. But his Pangea Ultima forms differently from Amasia or Novopangaea. Scotese believes subduction will start up on the west side of the Atlantic. The Mid-Atlantic Ridge is then eventually pulled into the Earth. The widening stops and the Atlantic begins to shrink.

Late Permian, 258 Ma

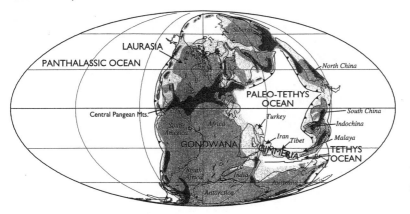

Paleogeographic maps by C. R. Scotese, PALEOMAP Project, University of Texas at Arlington (www.scotese.com)

Eocene, 50 Ma

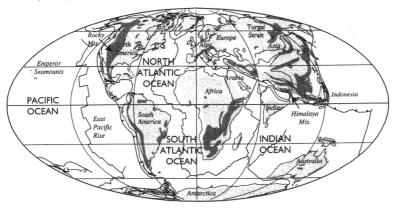

Scotese told reporters in 2000: 'Tens of millions of years later, the Americas would come smashing into the merged Euro-African continent, pushing up a new ridge of Himalaya-like mountains along the

Modern World

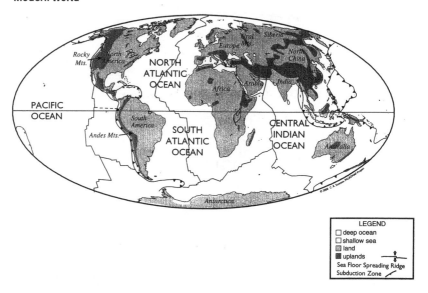

Late Permian, Eocene and today's world maps, showing the break-up of Pangaea according to modern research. © 2002, C. R. Scotese, Paleomap Project.

boundary. At that point, most of the world's landmass would be joined.' The result, however, is very different from Hoffman's Amasia or Livermore's Novopangaea. It looks like what it is: Pangaea reformed.

As a result of the news coverage generated by the NASA story, the name Pangea Ultima is now out there in the wider world in a way that Amasia never was. But there is something slightly wrong with its name. Scotese has called it 'ultima' because, as his website proclaims, 'it will be the last supercontinent to form'. But in reality the *next* supercontinent is just that. Whether Pangea Ultima, Amasia or Novopangaea, the next supercontinent will break up in turn and many other supercontinents will form again before the Earth is destroyed. Perhaps a better name might have been Pangaea Proxima.

Long, long ago in a galaxy far, far away

On a planet like ours, orbiting a sun that itself will one day die and bring geological time to an end, nothing is for ever. The deep time with which geologists conjure every day will – as our space explorer from Betelgeuse discovered – wipe away all traces of everything, including us. And when geological time does come to an end, the even deeper time of the cosmologist will erase all traces of everything.

Since the universe began, 13.7 billion years ago (plus or minus 0.2 billion), stars like our sun have been forming, burning and blowing up, their nuclear fusion furnaces making progressively heavier elements out of hydrogen, the simplest and most abundant atom. We and our rocky planet are made from substances composed of those heavy elements (carbon, oxygen, nitrogen, sulphur, iron, silicon) thanks to long-vanished stars. So even the immense lifespan of the Earth has a context in which it seems small. At this scale everything must go.

Think about what this means. Perhaps once, orbiting those lost stars that made the immortal atoms that build your body, there may have been some planets, perhaps like ours, on which there may have been life; life that may have become sentient, and that may have developed civilizations far beyond ours. We cannot possibly know this for sure; but there has been time, since the universe came into being, for it all to have come and gone – more than once. The only traces now left of those possible worlds may be those atoms in you, formed in the stellar explosions that banished whole histories to oblivion and wiped the slate clean. And one day, when there are no more days, that will be precisely all that's left of us too.

But not yet. Although Earth is about 4700 million years old at the moment, it is actually a galactic youngster. The universe was already 9000 million years old when the Earth started to form from cosmic dust and debris. Earth's lifespan is determined by the Sun, which will engulf our world in its death throes. So not only do we know how old Earth actually is, we also know (because we understand how long it will take the Sun to exhaust its fuel) how old the Earth will get. From this we can say that our planet has passed her middle age but still has a way to go: perhaps five or six times longer than the entire period that complex life has lived on her surface.

We began with a glimpse just a little way into our planet's (and perhaps our species') future, a mere 200 million years or so, by which time the next supercontinent will have formed. Apart from giving some sense of the immensity of the time with which we will be dealing, I hope it illustrates another central point of the supercontinent story; and that is how building imaginary worlds stirs passions.

Imagined worlds, both past and future, embody assumptions that affect our vision of ourselves, our past and future, our identity, our prospects. As human beings, our own species and what might happen to it lies at the heart of most of our thinking. The unsettling thing about the universe is the fact that within it our existence has no importance; and the unsettling thing about science is that it reflects that. The effect has come to be called the 'progressive dethronement' of Man. Science attempts to find out how things really are, rather than (for example) to frame myths to explain things away while at the same time flattering our vanity by putting humans at the centre of everything. When geology rebuilds the lost worlds of nature, the assumptions it employs put no weight upon human beings' mental comfort.

Building lost worlds, now scientifically the domain of geologists, has been with us a lot longer than geology, which as a scientific discipline is barely two centuries old. Lost worlds as an idea reach back to the earliest of our planet's explorers, who speculated about continents that may have once existed, or might still exist, somewhere, beyond or under the sea. Wondering what lies over the horizons of the deep oceans and deep time is no dry pursuit, through the centuries it has held the power to embody dreams, hopes and fears – of idyllic pasts and futures, of nationhood, myth and legend – even of God and divine purpose. And in the mêlée of history the two have often become confused as scientific ideas about possible 'lost worlds' have escaped the domain of science and taken on new life as myth.

Here be parrots

After 1492, travellers' tales began to feed rumours of lost continents. The great world map *Typus Orbis Terrarum*, published in Antwerp in 1570 by Abraham Ortelius (1527–98), fossilized many of these ideas into a kind of reality. As well as a host of fictitious islands in the South and North Atlantic, Ortelius depicted massive unknown continents covering the (then unvisited) North and South Poles. These were the first lost continents to be endorsed by something we might recognize as 'science'. And you can't miss them.

Ortelius was the son of an Antwerp merchant and started illustrating maps at the age of twenty. He was no traveller himself; he preferred the information to come to him. Dictionaries of scientific biography have traditionally been rather sniffy about the 'uncritical' Ortelius, because he collected all the information available to him about what people thought the world was like, and naturally much of it was wrong. Now, this fusion of the known with the asserted, the rumoured and the traditional, is precisely what is most fascinating about his 'theatre of the world'.

'*Terra septemtrionalis incognita*' ('Unknown northern land') it says across the top of the map, and '*Terra australis nondum cognita*' ('Southern land not yet known') across the bottom. These unknown polar lands, especially the southern one, almost dwarf the known continents of the world. Although *Terra australis* occupies the site of the then undiscovered continent of Antarctica, it is very much larger. Even though it is *incognita*, Ortelius does record a few details. Facing the Cape of Good Hope, for example, he writes '*Psittacorum regio*', and goes on to explain how, 'according to the Portuguese', this great southern continent is inhabited by giant parrots.

At one point, in a strange coincidence with geological reality, the parrot-infested *Terra australis* reaches out to touch South America at Tierra del Fuego. Antarctica was indeed once joined to South America across Drake Passage, the region of ocean floor studied by Roy Livermore. Antarctica only became today's frozen continent about thirty million years ago when that link was broken and the circumpolar current (which isolates Antarctica from the heat of the world ocean) finally shut the freezer door.

Ortelius did not invent these continents from nothing. The idea of a great southern continent began life as an ancient Greek notion of a 'counter-Earth' 'balancing' the known continents of the Northern Hemisphere, and named by Hipparchos of Rhodes (190–125 BC), who coined the term 'Antichthon' for it. Hipparchos even speculated that Sri Lanka might represent this southern continent's northernmost extremity, thus joining a long line of writers, from ancient Tamil poets to modern geologists, to embroil the southern extremities of the Indian subcontinent in stories of lost lands: real, imaginary and somewhere in between.

Ortelius's great map, and the *Thesaurus Geographicus* that he published six years later, start us off on the story of vanished supercontinents. It is barely ten years, in fact, since historians realized that Ortelius was also the first to speculate, from the fit of the opposing shores of the Atlantic, that this ocean may have arisen by the horizontal displacement of its bordering continents. Supercontinents and continental drift were born twins.

Not until 1994, in a paper in the British scientific journal *Nature* by a US historian of science, James Romm, did Ortelius finally get the credit he deserved. In the 1596 edition of his *Thesaurus Geographicus* Ortelius speculated about Plato's allegorical 'lost world' of Atlantis, which by that time was widely regarded as a piece of true history. He went on to make two scientific breakthroughs. He noted how the opposing shores of the Atlantic were congruent, and he then speculated about how some catastrophe might have separated them. He concluded that if Plato was to be regarded as accurate history, then his work should be reinterpreted in terms of lateral dislocation of the opposing continents, rather than subsidence.

They say the best place to hide information is in a library, and of all the books in a library the most secure are encyclopaedias. So it was that Ortelius's insight lay buried for four centuries as a single entry in a huge, outdated and unread work of reference.

2

ICE AT THE EQUATOR

*History warns us . . . that it is the customary fate of new truths
to begin as heresies and to end as superstitions.*

THOMAS HUXLEY, 1878

Bouverie Street is a short and rather drab offshoot of Fleet Street in
London. Number twenty-seven, which later was to become the office
of a newspaper and of its editor, Charles Dickens, was originally built
by William Blanford to house a small manufacturing business as well
as himself and his wife. She bore him two sons, William Thomas and
Henry Francis Blanford. Both became geologists and, like many
others of their generation, pursued their life's work in India, work
that would lead them to make the wild surmise that Southern
Hemisphere continents were once united.

William, born in 1832, was the elder by two years. He would live
twelve years longer than his younger brother, dying in 1905 covered
with scientific glory: Fellow of the Royal Society and President of the
Geological Society of London. And in a world obsessed with priority, it is William who is generally credited with being the first to notice
the striking geological similarities between rocks of the now widely
separated southern continents and draw attention to a conundrum
that would puzzle geologists and biologists for the best part of a hundred years.

As red spread over the world map, sons of England sought their fortunes – financial, military and scientific – among the Empire's furthest reaches. And as geologists were off studying the rocks of these distant lands, a new species of biologist (the biogeographer) began mapping the distribution of animals and plants across the known world. It was not long before scientific London began to receive reports of some odd patterns that demanded explanation.

These patterns, in the distribution of rock types, in the fossils they yielded to the hammer, and in the living things that grew above them, boiled down to two basic and very puzzling facts. Some things that were very similar to one another were being found much farther apart than their similarity would suggest possible; while other things, utterly different in character, often cropped up much closer together than their dissimilarity should demand. Every scientist who added another fact to this mounting body of evidence instinctively knew that this was important. But what was it telling them?

King coal

The *annus mirabilis* for the Blanford brothers was 1856. David Livingstone was returning triumphantly to Britain after his coast-to-coast exploration of the 'dark continent' of Africa. In Germany's Neander Valley the first fossils of another species of human being, *Homo neanderthalensis*, were being unearthed. And in India the Blanfords were surveying coal-bearing rocks in the eastern state of Orissa.

The brothers had arrived in India after being offered posts at the country's nascent Geological Survey by its Superintendent, Dr Thomas Oldham. Unfortunately, Oldham had neglected to tell anyone about either appointment, so, when the Blanfords landed in the late summer of 1855, nobody in Calcutta was expecting them. Oldham was away in the field. The Survey had no offices. They were stranded.

By one of those absurd pieces of chance that sometimes attend the fortunate traveller, amid the hurly-burly of one of the biggest cities on the planet, the hapless brothers ran into a fellow staff member of the Survey, William Theobald. There were no telegraphs and the posts were very slow, so it was December before they finally met their new boss on his return.

The Blanfords had not wasted their time, filling the intervening months with excursions to the Raniganj coalfield and the study of Hindustani. And less than another month elapsed before Oldham dispatched them, with Theobald, on their first proper job. They were sent to examine and report on a coalfield near Talchir (Talcher), some sixty miles north-west of Cuttack, an important provincial town in Orissa.

The Talcher coalfield, today part of a company called Mahanadi Coalfields, still boasts reserves of 35.78 billion tonnes. It feeds several local power plants operated by India's National Thermal Power Corporation, the country's biggest generating company. What the Blanfords found, as well as coal, was evidence that before it had been deposited, in what they presumed to be lush, steaming tropical swamps, India had apparently suffered a massive Ice Age.

The telltale deposits lay at the base of a two-kilometre-thick series of sedimentary rocks rich in coals and plant fossils. Underneath these coal-bearing rocks lay another unit, the Talchir Formation. This consisted mostly of sandstones and shales; but at its very base, lying on top of an eroded and grooved ancient land surface, was a highly unusual deposit. Many boulders, some larger than a man, lay embedded in a matrix of fine mudstone. The curious thing about this was the coincidence of huge boulders with fine mud. No beach, river or seabed accumulates a deposit like that. Apart from volcanic mudflows (which this was not, though many in years to come would claim it was) only one known natural agent was powerful enough to have moved the boulders and yet also was capable of depositing them together with fine mud: a glacier.

Glaciers are extremely powerful erosive agents, gouging out their U-shaped valleys and breaking up the rock walls into debris of all sizes from boulders as big as buses to rock dust as fine as flour. And when glaciers melt and retreat, all the material carried by the slowly moving river of ice is dumped together. The result is called, appropriately, boulder clay, or sometimes till. And a fossilized till, one turned into rock by age and pressure, is a tillite.

Looking more closely at the boulders, it is possible to see scratches that also betray the action of ice, which grinds each piece of debris over its neighbour with huge force. It is also likely that many of the boulders have travelled hundreds (even thousands) of kilometres, and a careful comparison with a geological map can help you work out the direction in which the ice moved.

But, as we shall see many times in the story of reconstructing supercontinents, it is not always easy to admit what the rocks are telling you when your mind refuses to believe it. This is what makes the Blanfords' conjecture so amazing today. The two young men returned to Calcutta and submitted their first report to their boss. In it they said they had discovered unequivocal evidence that ice sheets once covered what is now tropical India. The intellectual toughness that this betrays is matched only by the fact that their boss appears to have believed them.

By no means everyone was convinced. It simply seemed impossible. Even as late as 1877 the recorded discussions of a paper given to the Geological Society of London by the younger Blanford are full of disbelief. But by that time the same boulder bed had been traced beyond Talchir throughout a wide area of Bengal and the Central Provinces. By 1886, when the elder Blanford read another paper at the Society, boulders bearing those unequivocal scratches had been found. There was no longer any room for serious doubt. The Blanfords had been right all along. Somehow, all those millions of years ago, there had indeed been sea-level ice at twenty degrees of latitude.

Explaining how this could be so became no easier in the decades following the discovery. A similar boulder bed (now named the Dwyka Formation) had been discovered at the bottom of a similar succession of rocks bearing similar fossils in South Africa. Henry Blanford had already suggested that these were the same as the glacial deposits he and his brother had first seen in India. Moreover, as early as 1861, the frequently absent Mr Oldham had tentatively correlated the Talchir boulder bed with one he had seen on a visit to Australia. These, too, had cropped up all over the country, from Queensland in the north through Sydney to Wollongong in the south, and in the Blue Mountains in the west. This troublesome glaciation, which seemed impossible enough even when confined to India, now appeared to have spanned the Equator and covered half the globe.

Further investigations were throwing up bigger questions than they were answering. Could the Earth perhaps even have tilted on its axis, as Oldham had suggested? And what was the precise age of the glaciation? The fossils of Southern Hemisphere rocks were being described bit by bit; but they were very different from the better-known fossils of the Northern Hemisphere. It was like trying to navigate by the southern stars knowing only the boreal constellations. Maybe this glaciation had coincided with one of the mass extinctions that had already been recognized in the fossil record. Was there a connection? Nobody could be sure, but everyone had an opinion about why so many similar rocks, with their identical fossils, were found so widely scattered across the continents of the Southern Hemisphere, and how a glaciation could occur at the Equator.

Looking back at the lives of geologists from these heroic days makes one doubt that Victorians were made from the same stuff as we are. William Blanford worked for twenty-seven years in the Indian Survey, during which time he travelled and mapped widely, not only within the subcontinent but through Afghanistan and what was then Abyssinia and Persia. He retired from the Geological Survey at fifty and bought a house in Kensington. He married, settled into London's scientific scene and began his second career.

W. T. Blanford had already been elected a Fellow of the Royal Society in 1874. Before him now lay many years of office-bearing, not only at the Geological Society of London but also at the Royal Geographical Society and the Zoological Society of London. During his travels Blanford had noticed many things that puzzled him about the living world, and his retirement offered him the chance to complement his geological work by researching more fully the distribution of species across the subcontinent that had been the main interest of his life.

In 1890 Blanford wrote the following words, which have since turned out to be even truer than he knew: 'all who recognise how intimately the story of the Earth is bound up with that of its inhabitants will have little doubt that the present distribution of animals and plants is of the highest geological importance, and that the existence of particular forms of living beings in continents and islands is the result and the record of the history of those areas and *of their connexions with each other* [italics added].'

Presidential addresses to the Geological Society have often been long, but few rival the fifty-four pages of the Society's *Quarterly Journal* that are occupied by Blanford's second; and few Presidents (even including that of zoological luminary Thomas Henry Huxley, exactly twenty years before) had quite the nerve to deliver one so completely non-geological. But this was probably deliberate. Geologists had to be made to take biogeography seriously.

Grand tours

Biogeography had come into its own as the great trading empires of the West had fanned out across the globe, taking their naturalists with them. A small army of botanizers and hunters, including many eminent scientists, set off to find, draw, paint, capture, skin, stuff and in some cases send back their captives alive for the fascination of the London public. They also returned home with new ideas, ideas that would break first on London and quickly overwhelm the world with their significance.

Charles Darwin had his eyes opened as the gentleman-naturalist companion to the captain of the *Beagle*. For the man who would be his champion, the less genteel Thomas Henry Huxley, the experience was to be a spell in Her Majesty's Navy aboard a leaky frigate called HMS *Rattlesnake*. As the young ship's surgeon on his first job, Huxley was forced to inhabit a cramped berth, frequently awash, and to watch his shipmates die of injuries and fevers he had no medicine to cure. Science was feeling the spur of Empire.

Another of the great biogeographers of the nineteenth century, Philip Lutley Sclater (1829–1913), sailed for the Americas in 1856. In the course of his travels over the next decades he was not only to cover most of the USA but also many of the continents linked geologically by those boulder beds and plant fossils (including Argentina, Australia and India) and, crucially, the lands of the Malay Archipelago. He was also the first modern scientist not only to propose but also to *name* a hypothetical vanished supercontinent.

After he returned to London, Sclater's career was worthy and long. For forty-four years he held the influential post of Secretary to the Zoological Society of London. But he made his mark as an original thinker just two years after leaving on his first great trip, by triumphantly dividing the living world into six great realms defined by distinctive assemblages of animals, realms that are still recognized today. As an ornithologist first and foremost, Sclater began with the birds, Class Aves as zoologists have it.

His paper to the Linnean Society on birds' global geographical distribution became an instant classic. He soon began to incorporate other animals into his scheme, and twenty years later wrote a review in the popular and influential monthly periodical *Nineteenth Century* in which he recounted the curious distribution patterns of the lemur.

Lemurs are primates, grouped together broadly as Lemuriformes. They are less closely related to humans than monkeys or apes are, and much more ancient in evolutionary terms. As defined today, they are limited to Madagascar and the Comores, where they have diversified into fifty-five species and subspecies, including the mouse and dwarf lemurs, the true lemurs, sportive lemurs, the woolly lemurs and the ghostly aye-aye, the nocturnal grub-eater with one modified long finger that it uses to winkle its prey out of wood.

At the time Sclater was writing he used a broader classification of lemurs than our modern one, including similar, related primate groups found in southern India and Sri Lanka and the scattered islands of South-East Asia. The pattern he found recalled the one geologists were puzzling over with their boulder beds and plant fossils. Lemur-like primates had a scattered distribution that didn't make sense. One could believe, for example, that lemurs might cross the strait between Africa and Madagascar on rafts. But was it likely that these little creatures (or their common ancestors) could cross all the way to Sri Lanka and the Malay Archipelago, traversing the Equator and thousands of miles of ocean, or float there on rafts washed from distant shores?

It was 1864 when Sclater first published his proposed solution to the problem, in a paper called 'The Mammals of Madagascar' in the *Quarterly Journal of Science*. He wrote:

> The anomalies of the Mammal fauna of Madagascar can best be explained by supposing that . . . a large continent occupied parts of the Atlantic and Indian Oceans . . . that this continent was broken up into islands, of which some have become amalgamated with . . . Africa, some . . . with what is now Asia; and that in Madagascar and the Mascarene Islands we have existing relics of this great continent, for which . . . I should propose the name *Lemuria*!

Lemurs could not have swum between these far-distant places – they must have walked – so there must have been land.

Sclater was not the first scientist to suggest a lost southern continent to explain biological (or geological) anomalies. Geoffroy St Hilaire (1772–1844), a French natural historian, had suggested one in the 1840s, having also noticed the peculiarities and Indian affinities of Madagascar's animals. What Sclater did, however, was give his creation life by *naming* it. This enabled his 'Lemuria' to escape from the world of science and enter common knowledge. Lemuria, a place that never really existed, began to rise to the status of an Atlantis, to enter into myth and inhabit that same strange hinterland of half-truth. By the 1880s the idea of Lemuria had become well entrenched in the literature. The ghost had entered the machine.

Toeing the line

Another travelling naturalist whose story intersects with ours at this point was Alfred Russel Wallace (1823–1913). Though later critical of invoking sunken continents as a way of explaining biogeographic provinces, at first he eagerly adopted Sclater's Lemuria in his writings, believing that the giant vanished continent must have extended from 'West Africa to Burma . . . South China and the Celebes'.

Wallace is mainly remembered today for hitting upon natural

selection as the driving force of evolution independently of Darwin. This is the great Welsh naturalist's other great claim to fame. Unlike Sclater and the geologists (who noticed things that were *too far apart to be so similar*), Wallace's breakthrough came when he began to notice certain animal species that were *too different to be so close together*.

As you sit in a small beachside café in Padangbai, Bali, waiting for the Lombok ferry to arrive at the jetty on the southern headland, you look out at the palm-covered arms of land that enclose the bay, the fishing canoes drawn up on the sand, their great outriggers and prows painted like crocodile jaws, and you can make out your destination quite easily. Mysterious Lombok stands, densely wooded, against a pale horizon. Its classic cone shape tells you that you are looking at Mount Rinjani, Lombok's 3727-metre volcano. In fact, though, you are looking only at its top half, and missing completely all of the southern, low-lying part of the island. It lies below the horizon, hidden by the curvature of the Earth.

The guidebook you are reading will tell you how different Lombok will feel from easygoing, relatively well-to-do Bali, with its colourful animistic religion, its devotion to flowers, its picturesque processions, representational art, rich music and dance culture and toleration of alcohol. For a start Lombok is much poorer. Even quite recently people have died of starvation there after bad harvests. And despite pockets of Balinese influence in the west, a form of Islam dominates the island.

It is tempting to draw a parallel between these cultural divides and the next thing your guidebook will tell you, which is that the short twenty-five kilometres of water you will soon be crossing separate two completely different animal realms: one Australasian, the other Eurasian. And you will be told that the divide is named after Alfred Russel Wallace, co-founder of evolution theory.

Wallace spent much of his life in or around this archipelago (some of it engaged in catching a live bird of paradise for Philip Sclater at the Zoological Society of London). The 'line' he set out bisected the archipelago, running south of Mindanao, between Borneo and Kalimantan to the north of where you sit in Padangbai, and threading, like cotton through a needle's eye, along the Lombok Strait. Like Sclater and his great realms, Wallace based his first observations primarily on bird species (mainly parrots, though not giant ones).

The Lombok Strait, your guidebook might also say, is a 300-metre-deep channel with some of the strongest currents in the world, which is why the ferry takes four hours. But despite this fact, the Wallace line is not actually as sharp as they would have you believe. Depending on which animals you look at, you can draw many different lines through the scattered islands of the region. Sclater, for example, working with his son William, defined a different line much further to the East, dividing Celebes and Timor from Irian Jaya.

There are many other, less famous dividing lines, and even two 'Wallace lines', for the great man allowed himself second thoughts in 1910. Weber's line (1904) is based on the distribution of freshwater fish, and farthest east of all lies Lydekker's line (1896), hugging the edge of the Australian coastal shelf. It all depends on which animal you take as your most important marker. Different species have different abilities when it comes to dispersing themselves. Certain birds, for example, are more able to cross deep salt-water straits with strong currents than amphibians or freshwater fish.

Wallace never truly understood what his faunal break meant nor why it was so abrupt on a global scale; but part of the answer lies in the depth of the channels that separate the islands. Even during the last Ice Age global falls in sea level never much exceeded 125 metres and so never exposed the bottom of the Lombok Strait. A land bridge was never established. But if the seas around Bali had been shallower, or the last Ice Age more severe, the whole modern pattern of animal distribution would have been radically different. Today, instead of choosing any particular line, biogeographers acknowledge a zone – the region between Wallace's line in the west and Lydekker's line in the east – and have dubbed it Wallacea: a broad buffer between two of Philip Sclater's great faunal realms.

Until geologists found a mechanism that could make neighbours of such different animal assemblages with such different evolutionary histories, the origins of Wallacea would remain another mystery. And, unlike those animal assemblages that were too far apart to be so similar, it could not be explained by suggesting that large tracts of ancient continent had disappeared under the waves, stranding the lemurs and all their friends on now-distant shores.

Weird science

The ring-tailed lemur was first described and named scientifically in 1758 as *Lemur catta* by the father of classification, Carl von Linné (or Carolus Linnaeus, to use his Latinized name), the Swedish taxonomist (1707–78) who gave his name to the Linnean Society of London, where Darwin and Wallace's joint paper on natural selection was first presented in July 1858. All Linnaeus knew about these nocturnal primates was that they had ghostly faces, trimmed with haloes of white fur, big, forward-pointing, cat-like reflective eyes (that stared unnervingly at you out of the jungle) and that they made ghostly cries in the night that chilled the blood. He may also have known that local legends held lemurs to be the souls of ancestors. And so he chose a name that seemed appropriate, deriving it from *lemures*, the name used by the Romans for ancestral ghosts. (In pagan Rome these household ghosts had their special days (9, 11 and 13 May of the Julian calendar), days that became known as Lemuria.)

So when Philip Sclater brought the word 'Lemuria' back from the dead to denote a ghostly, vanished continent he unwittingly linked it to the occult. The circularity of this etymological accident may seem poetically satisfying and appropriate, but the name does almost seem to have brought a curse. Mythical lands attract strange settlers.

The astronomer Percival Lowell filled his fanciful reconstructions of the surface of Mars (1896) with cities and irrigation canals, yet he still forms part of the history of legitimate astronomy. But, in the days before people invented myths about aliens from other planets, they invented races that peopled long-vanished terrestrial continents. In the same way the fantasies of mystics and the age-old notion of foundering continents grade imperceptibly with the early science of the supercontinent story and prevailing theories about how the crust of the Earth actually moves.

The energetic and wildly enthusiastic German biologist Ernst Haeckel (1834–1919) got hold of the idea of Lemuria and suggested that Sclater's continent had also been the cradle of human evolution. In the *Natürliche Schöpfungsgeschichte* of 1868 (translated into English as *The History of Creation*, 1876) he published a map showing the radiation of humans across the entire globe, with all the myriad branches converging on the hypothetical lost continent.

This immediately raised the political stakes, as anthropology always does, because it closely affects people's assumptions about themselves. The origin of humans is a charged science, and more interesting to most people than the origin of lemurs. Sure enough, it brought Lemuria to a wider audience, especially as Haeckel specifically linked the place to myth by subtitling the continent, in brackets, '*Paradies*' (Paradise).

Here are two widely separated examples of how Lemuria caught on. In 1876, the year Haeckel's English translation was published, Lemuria made an appearance in Friedrich Engels's *The Part Played by Labour in the Transition from Ape to Man*. On his opening page Engels wrote: 'Many hundreds of thousands of years ago, during an epoch . . . which geologists call the Tertiary . . . a particularly highly-developed species of anthropoid apes lived somewhere in the tropical zone – probably on a great continent that has now sunk to the bottom of the Indian Ocean.' Even H. G. Wells included a reference to Lemuria in his *Outline of History* (1932), airily speculating that the 'nursery' of humankind 'may have been where now the Indian Ocean stands'. Lemuria thus became a textbook fact and, in the way of such things, remained so many years after most legitimate science had abandoned the whole idea.

Crucially for what followed, Lemuria also became a textbook fact in India, through some ancillary work carried out by the Blanford brothers. In 1873 Henry Blanford wrote a schoolbook of physical geography in which he told his Indian readers that their continent had once been linked to Africa and that this link had been sundered by enormous outbursts of volcanic activity. Six years later William Blanford's Indian Geological Survey published its *Manual*, describing the country's geology. Here the case for a former geological link between India and southern Africa was clearly made. These scientific pronouncements found particular resonance among certain peoples in south-eastern India, the Tamils.

Lemuria and Katalakōl

In 1974 Léopold-Sédar Senghor (1906–2001), poet and founding President of Senegal from 1960 to 1980, addressed the International Institute of Tamil Studies in Madras. He mentioned 'the cradle of mankind' and located it in the Indian Ocean just as H. G. Wells had done. He then went on to bring the date of its destruction under the waters of the Indian Ocean forward to the New Stone Age and

suggested that those places traditionally thought of in the West as
'cradles of civilization', Egypt and Mesopotamia, derived their
knowledge from the peoples who had fled this cataclysm.
Archaeologists, he said, should have the chance to explore the depths
of the seas, 'to discover old lithic industries or human skeleton fossils
in the area stretching from East Africa to Southern India'. Senghor
was a member of the Académie Française and the author of many
books of poetry and political philosophy; but from a scientific point
of view this was a very odd thing to believe in 1974.

In 1981, as part of the Fifth International Conference of Tamil
Studies in Madurai, a documentary film was screened. Made with the
personal backing of Chief Minister M. G. Ramachandran and paid
for by the government of Tamilnadu, this curious film traced the
origin of Tamil language and literature to its most distant past on
Kumarikkantam, a mythical lost homeland, redolent in Tamil
mythology.

Tamilnadu, the Indian state that stretches from the southern tip of
India at Kanyakumari to Tiruvallur on the Bay of Bengal, is home to
some fifty-six million people who speak Tamil, one of twenty-six languages spoken mostly in southern India and Sri Lanka and
collectively called Dravidian. While many show varying degrees of
Sanskrit influence, Tamil is the purest of all.

Tamil has a rich ancient literature and, crucially for this story, one
myth in particular that cites a terrible flood that catastrophically swallowed up the Tamils' formerly much more extensive homeland,
destroying much ancient lore, civilization and majesty, and leaving the
people with that small remnant of peninsular India which they now
inhabit. Tamil literature refers to this inundation as Katalakōl.

The scholar Sumathi Ramaswamy dates the appearance of Sclater's
Lemuria in modern Tamil writings to the late 1890s, when Tamil
authors first made the connection between it and the mythic home-

land of Kumarikkantam, sometimes even referring to this lost continent as 'Ilemuriakkantam'. Modern science was being called upon to lend a picturesque Tamil myth the aura of literal truth.

The result of this process (which continues to this day) has been to set up a curious dichotomy. While academic historians in Tamilnadu do not necessarily believe in the literal truth of Katalakōl any more than their geologists might, among ordinary folk in Tamilnadu there is a commonly held belief that Western science 'backs up' the claims of their mythology. This continues to be implied, more or less overtly, in Tamil Studies.

Citing Sclater, Blanford and Haeckel, Tamil 'devotees' (as Ramaswamy terms them) thus imply (and in many cases openly claim) that the Tamil people are the ancestors of humanity and that their language is not only more ancient than Sanskrit (which is true), but is also the mother of all Dravidian tongues and even the oldest language of mankind. To put this into a more familiar English context, this is rather as if students of English were taught Arthurian texts with the clear implication that all that 'sword and sorcery' really happened. As Ramaswamy writes, this belief serves a political purpose: 'The collective yearning for an unreclaimable past plenitude holds together a people in exile otherwise riven apart by caste, class and religious differences.'

However, there is a fly in the soothing ointment. The lost supercontinent of Lemuria never existed. The claim that it affirms the literal truth of the Tamil flood myth, via outmoded Western science, is an empty one. Science has moved on. By tying themselves to an unsustainable insistence on literal truth, Tamil's devotees align themselves with outdated science, while at the same time depriving themselves of a more positive (overtly acknowledged) mythology. Myths, after all, can contain truths other than literal ones, truths that can be more fruitful. As Ramaswamy writes: 'A mongrel formation, neither pure fantasy nor respectable history, Tamil labours of loss are vulnerable to disavowal and dismissal both as fantasy and as history.'

Those familiar with the argumentation of so-called 'creation science' will find this philosophical bind oddly familiar. By insisting on the literal truth of the creation myth told in the Old Testament and by vainly looking for evidence of the supernatural among the things of this world, young-Earth creationists saddle themselves with a similarly impoverished mix. What they espouse is both bad science and bad religion, demonstrating nothing more than ignorance on the one hand and lack of faith on the other.

In Tamilnadu the conflict between modern science and an ancient myth propped up by the trappings of science (variously outmoded, selectively quoted or fraudulently misrepresented) rarely becomes apparent. One senses that those who know of the conflict (rather than purported accord) between myth and modern science see little purpose in challenging a widely believed misconception that does little practical harm.

However, every now and then political pronouncements can force the issue. In 1971 the Tamilnadu government decided to assemble a panel of scholars to 'write the history of Tamilakam from ancient times to the present'. The education minister, R. Nedunceliyan, observed at the time: 'When we say history, we mean from . . . the time of Lemuria that was seized by the ocean.' That sort of thing can (and does) make scholars uncomfortable.

As Ramaswamy notes, this Tamil belief remains 'eccentric' within the culture of Tamilnadu. The spurious linkage of myth to science is confined to courses about Tamil, taught through the medium of Tamil, using textbooks produced by local publishers and the government-run textbook society, and catering predominantly for those (mainly less advantaged) people attending state schools. And so the belief persists, a hand-me-down textbook fact, immortal but (mostly) invisible.

Wave of death

Facing Sri Lanka across Palk Bay is the tip of India. Once dubbed
Cape Comorin, it is now known by its original name of
Kanyakumari. In the 1950s it became the subject of a concerted political effort to ensure that it was incorporated into the emergent Tamil
state, which took its present form under the Madras presidency in
1969. 'Kumari' is a powerful name in Tamil. It is the name of the
mythical lost continent itself. It is also the name of one of its supposed mountains, and also of a major river that supposedly crossed
it. Kanyakumari is the only surviving real place still to bear this heavily loaded name; a last bastion against the cruel sea that tore the
Tamil homeland from its people in Katalakōl. It plays the role of 'a
vestige of the vanished'.

Just a few hundred metres offshore is a tiny islet on which the
Vivekananda Memorial now stands. It is composed of a rock called
charnockite, a rock first identified in the tombstone of the first
governor of Calcutta, Job Charnock (d. 1693), which was formed
deep in the Earth's crust 550 million years ago as two continental
masses fused together. Geology does not admit to having sacred
sites, but if it did, this would be one of them. For reasons that we
shall explore later, geologists refer to this island as 'Gondwana
Junction'.

Like many of the world's sacred sites, the islet is sacred to more
than one group. On 26 December 2004, at nine o'clock in the morning, about five hundred pilgrims, mostly Indian, crowded on to the
rock to stand at the Vivekananda Memorial and see the sun rise.
Although they would not number among them, before that day was
out, over 200,000 people all around the Indian Ocean would be dead.
Many had already died; waves of destruction and death were spreading outwards, triggered by the biggest earthquake for fifty years,
caused by the very processes that are forming the next supercontinent.

An event that has certainly happened before, and will just as certainly happen again, was about to overwhelm Tamilnadu.

Before Boxing Day 2004 the most recent Global Geophysical Event, which by definition affects (in some way) people on every continent, had been the eruption of Krakatoa in 1883. The Tsunami likewise had the effect of drawing the world together, but it also united Earth scientists in frustration that their knowledge had not been used to full effect, for example, to set up early-warning systems that had existed in the Pacific since the end of the Second World War. It also explained two things. It explained why Tamil people have an embedded myth of a dangerous and land-hungry sea that snatches life away in a catastrophic Katalakōl. And for me it explained why the seemingly abstruse subject of the Earth's Supercontinent Cycle matters to everyone, everywhere.

But now is not the moment.

3

QUEENS OF MU

'Some kind of legend from way back, which no one seriously
believes in. Bit like Atlantis on Earth.'
DOUGLAS ADAMS, *THE HITCHHIKER'S GUIDE TO THE GALAXY*

Philip Sclater's hypothetical lost supercontinent, invoked to explain
the scattered distribution of lemurs, was haunted right from the
outset. But of all the strange settlers Lemuria attracted, none was
stranger (or had wider influence) than Helena Petrovna Hahn.

Hahn (1831–91) was born in the southern Ukrainian city now
known as Dnipropetrovsk but which was then known as
Ekaterinoslav; she was the daughter of Pyotr Alexeyevich von Hahn,
an army colonel, and his novelist wife Elena Fadeev. Elena, who had
earned herself the literary sobriquet of 'the Russian George Sand',
died when her daughter was only eleven. Although the family had
moved around considerably, as army families do, her father was
unable to take little Helena with him after her mother's death so Pyotr
Alexeyevich farmed her out to her maternal grandmother, Elena
Pavlovna de Fadeev, she was nobly born, a well-known botanist, and
another formidable character.

Helena Petrovna was to grow up to be one of the strangest women
of the nineteenth century, going from sweatshop worker to bareback
rider, professional pianist and finally co-founder and guru of a pop-

ular and once-influential new religion called Theosophy. Helena Petrovna was the first of the New Agers, and she derived the name by which the world knows her today from her first husband, General Nikifor Vassiliyevich Blavatsky.

Escaping from the General soon after the wedding by breaking a candlestick over his head and fleeing on horseback to Constantinople, Madame Blavatsky – after another very short marriage – set off to travel the world, ending up in 1873 in New York, where she set up as a medium. There she teamed up with Henry Steel Olcott (a lawyer who left his family for her) and others and founded the Theosophical Society, a new religion combining aspects of Hinduism and Buddhism. This new creed, she claimed, had come to her in a 'secret doctrine' passed down from an ancient brotherhood. Unlike those of the Rosicrucians and Freemasons, Blavatsky's ancient brothers derived from Eastern rather than Western sources. And in common with many subsequent New Agers, Blavatsky claimed that her so-called Akashic Wisdom was consistent with science, and especially the then fashionable new science of evolutionary biology. This was a remarkable claim, since the scientific idea she most hated was the one that humans had evolved from apes. Madame Blavatsky had her own ideas about that and set her own distinctive account of human origins on landmasses that no longer existed. Lemuria, coming as it did with impeccable scientific credentials, fitted the bill perfectly, just as it had for Tamils.

Blavatsky had moved from America to India in 1879, and in 1882 she passed a number of letters from her late Master, Koot Hoomi Lal Sing, to an Anglo-Indian newspaper. (Graphologists later determined that she wrote them herself.) The cosmology contained in them was based on the number seven: seven planes of existence, roots of humanity, cycles of evolution and reincarnation. This scheme formed the basis for her book *The Secret Doctrine,* which became the main text of the Theosophical movement.

Before she could finish this opus, however, Blavatsky was hounded out of India. Two of her staff, Alexis and Emma Coulomb (who may well have been put up to it by Christian missionaries), threatened to expose her mystic feats as trickery, and Blavatsky returned to Europe, where she completed *The Secret Doctrine* in 1888. It ultimately derived, she wrote, from a 'lost' work called *The Stanzas of Dyzan*. According to these, modern humans were the fifth of the seven 'root races'. The third race had inhabited the lost supercontinent of Lemuria, bandy-legged, egg-laying hermaphrodites, some of whom had eyes in the back of their heads and four arms (though perhaps not both at once).

The Lemurians had, according to Blavatsky, lived alongside dinosaurs. As if this was not exciting enough, they also discovered sex. This turned out to be A Bad Idea (for the Lemurians) because it was the trigger, Blavatsky believed, for the destruction of their continent. Their surviving offspring (the fourth 'root race') were the Atlanteans. It was they who wrote the *Stanzas* and who gave rise to the fifth race, namely us. Modern humans would eventually give way to the sixth and seventh races, who would inhabit North and South America respectively.

Blavatsky died in London in May 1891 from a chronic kidney ailment aggravated by a bout of influenza, and was cremated at Woking cemetery. Rather like Lemuria, the movement she founded soon split up and sank in schism and recrimination, never maintaining the following it commanded while its high priestess was alive. (It is estimated to have peaked at about 100,000 worldwide and is known to have included several influential and otherwise apparently sane people.)

Theosophy, pioneer of a genre, lives on, as does its conception of Lemuria, though largely on the ethereal plane of the World Wide Web.

Mystic Mu

The Indian Ocean had its Lemuria, and the Atlantic had of course its Atlantis. But what of the largest ocean of all, the last surviving remnant of Panthalassa? The potential financial rewards for this kind of work are great, as Blavatsky had shown; and as any scientific fraud or unscrupulous journalist knows, it is a lot quicker to make things up than find things out. Crucially too, the Pacific is the closest ocean to California, the best place in the world to found new religions. Madame Blavatsky herself recognized this, and in her later writings began edging her Lemuria out of the distant Indian Ocean and into the Pacific for this sound business reason. Yet despite her tweakings, the Pacific Ocean still represented a huge vacant lot to the would-be supercontinent maker, and before long one was duly 'discovered'. The odd thing is, by the time its name broke upon the public in the twentieth century, it had already existed in the minds of (some) men for centuries.

Mu is perhaps the maddest of all imaginary lost continents. Its origins, however, lie not with the sciences of zoology, botany or geology, but with archaeology; and the very unscientific analysis of some very ancient writings.

Its existence was first proposed by one Charles Etienne, Abbé Brasseur de Bourbourg (1814–74). The Abbé travelled much of Europe and Central and South America in the service of the Catholic Church, and apart from missionary zeal his main life interest lay in the ethnography of the native peoples of America. In his later years he became convinced of pre-Columbian connections between American and Eastern races, connections for which the existence of the Pacific Ocean constituted something of a geographical snag.

The Mayans left very few written documents, and deciphering them has always presented acute difficulties in the absence of any equivalent of the Rosetta Stone that offers the linguist parallel texts of which at least one is known. Nothing daunted, the Abbé set about the task of reading the Troano Codex. This codex consisted of half of one of three surviving Mayan manuscripts, and it is now part of what is today known as the Madrid Codex. In his reading he thought he discovered references to a sunken land by the name of Mu, and leapt at the idea because it solved his ethnographical problems by bridging the Pacific. So Mu started life, rather like Lemuria, as a means of explaining a distribution pattern – only this time, of people.

The Abbé's references were next picked up by a widely (though uncritically) read Philadelphia lawyer and Minnesota congressman, Ignatius Donnelly (1831–1901), author of *Atlantis: the Antediluvian World* (1882). Unhappily for Donnelly, his literary judgement failed him over the de Bourbourg 'translation' of the Troano Codex on which he, the Abbé and the supposed continent of Mu depended. For the translation was, in fact, nothing of the sort.

De Bourbourg had 'interpreted' the Codex, having himself discovered a 'Mayan alphabet' devised by a Spanish monk by the name of Diego da Landa. Arriving in America with the Conquistadores, da Landa was among the first scholars to come across the vivid pictograms of the Mayan people. His so-called alphabet was nonsense; the Mayan writing system was not letter-based at all.

The Abbé's Troano Codex 'translation', which supposedly described (in highly elliptical terms) some great volcanic catastrophe, was nothing more than a figment of the Abbé's fevered imagination, spurred on by the application of da Landa's bogus alphabet. And, crucially for this story, during the process of his creative decipherment de Bourbourg came across two pictograms that he could not at first identify. Thinking, though, that they bore a slight resemblance to the symbols that da Landa asserted to be the Mayan equivalents of the letters M and U, the Abbé duly discovered the name of the ill-fated continent. Thus was Mu born.

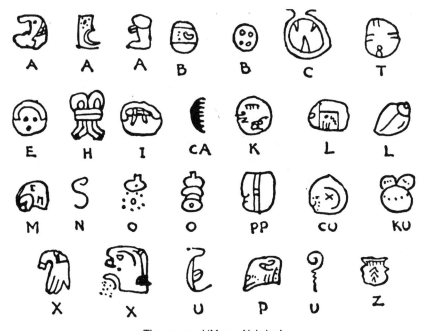

The supposed 'Mayan Alphabet'.

Back in Washington, unaware of how rotten its foundations were, Ignatius Donnelly took the Mu story on. In his book he linked this entirely bogus Mayan legend with Plato's allegorical Atlantis and went on to speculate about how this connection might shed light on archaeological links between the Mayan and other civilizations. And there the Mu legend paused, until one Colonel Churchward picked it up and moved it back into the Pacific.

Colonel James Churchward (1851–1936) stares winningly out of his picture like a cross between Colonel Sanders and a travelling medicine man peddling potions in a Hollywood Western. He sports a rakish goatee and moustache, and wears a large rose in his left lapel. Although he had written a book before, namely *A Big Game and*

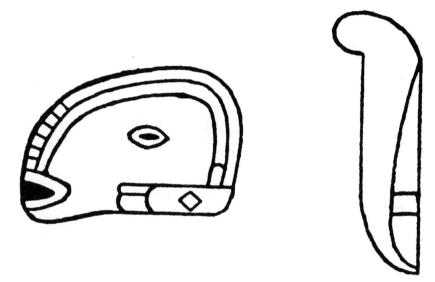

The two ideograms thought, from tenuous supposed similarities to characters
in the supposed alphabet, to represent the letters M and U.

Fishing Guide to North-Eastern Maine, this gasconading English
émigré shot to literary success rather late in life with his colourful
accounts of a huge continent lost below the Pacific.

The Lost Continent of Mu (1926) set out Churchward's claim to
have discovered the tale of Mu and its destruction in mysterious
ancient texts. He claimed the continent had sunk about 60,000 years
ago and that Easter Island, Hawaii, Tahiti and a few other Pacific
islands were its last remnants. This information he gleaned from the
'Naacal Tablets', having himself been taught the Naacal language by
a Hindu priest in India in 1866. (Churchward's military rank was said
to have been gained in the British Army in India, but this too is
unconfirmed.) As well as the tablets of Naacal, Churchward gleaned
information from a different set of tablets found in Mexico by one

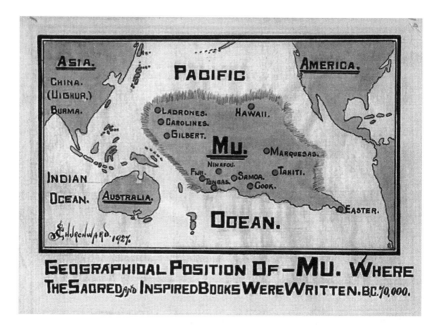

The lost continent of Mu as envisaged by Churchward.

William Niven, who is variously described as a geologist and engineer. No one else has ever seen these tablets either.

Churchward held that the first humans had appeared two million years ago on Mu. Modern humans were, he believed, all descended from the survivors of Mu's cataclysmic destruction, brought about by the explosion of the 'gas belts' on which it rested. Churchward followed his first book with four more: *The Children of Mu, The Sacred Symbols of Mu, The Cosmic Forces of Mu* and *The Second Book of the Cosmic Forces of Mu.*

One would think that after five volumes of elaboration (all of which are now back in print in America) Churchward's might have proved the last written words on the subject. But as recently as 1970 yet another book appeared, *Mu Revealed* by Tony Earll. This claimed to be the diary of a boy called Kland who, according to Earll, moved to Mexico in 21,000 BC but was unlucky enough to meet with an earthquake and get his scrolls trapped in a collapsing temple. Then in 1959, so the story goes, archaeologist Reedson Hurdlop excavated the temple. He found the scrolls and discovered that they not only supported Churchward's Mu hypothesis but provided even more information about the lost continent and its people.

Except all this was also fiction. Crossword enthusiasts may have noticed in passing that 'Tony Earll' is an anagram of 'not really' and 'Reedson Hurdlop' of 'Rudolph Rednose'. *Mu Revealed* was, in fact, the first novel by another émigré Englishman, the TV scriptwriter, self-styled witch and occult author Raymond Buckland (b. 1934). In the same year that he published *Mu Revealed*, Buckland also released (under his own name) *Witchcraft Ancient and Modern* and *Practical Candleburning Rituals*.

It is probably true to say that there is no stretch of land too miserable, too mean, or even too imaginary, that someone will not wish to be the king of it. For twenty years or so, beginning in 1933, the Office of the Geographer of the US State Department carried on a correspondence with a number of people concerning some alleged islands off Panama. One Mrs Gertrude Norris Meeker wrote in 1954 (on headed notepaper declaring her to be the Governor General of the Government of Atlantis and Lemuria) to point out that since 1943 a group of islands 200 miles south-west of Florida and just eight degrees north of the Equator had been the 'Private Dynasty or Principality . . . named "Atlantis Kaj Lemuria"'. 'Any trespassing on these islands or Island Empire is a prison offense,' the letter ended darkly.

The Department's geographical adviser, Sophia A. Saucerman, responded that the USA did not recognize such a state. In reply Mrs Meeker presented a detailed history of the Principality, involving a

Danish seaman called John Mott who in 1917, not wishing to return to a war-torn Europe, took possession of the place and founded the dynasty to which Mrs Meeker belonged.

In 1957 an official inquiry was set up 'to make a determination as to the reality of the Mu Group in the Pacific Ocean', as a result of which the Office replied to Bell that it did not believe these islands existed – and nor did it believe that anyone else believed it either. But the persistent Mrs Meeker then succeeded in persuading a US Congressman, Craig Hosmer, to take a hand in her affairs. In 1958 he wrote to the Geographer pointing out that, if her plans worked out, Queen Meeker of Mu might be a good source of trade. The Congressman's letter stimulated a swift reply. Three days later the Department pronounced itself definitively unaware of the existence of any such island empire: 'However, the Geographer of this Department is most willing to make a geographical study of this matter . . .'

'The file ends with this letter,' writes Sumathi Ramaswamy. The Geographer's kind offer to conduct research in the South Pacific was not taken up.

Sunken lands

By the end of the nineteenth century geologists and biogeographers had found out much about the rocks, fossils, animals and plants of the world, notably on the previously little-known southern continents. They had found sequences of rocks that looked so similar, it was incredible that they were now so far apart. Equally improbable was the fact that these rock sequences began with boulder beds, which suggested there had been a massive glaciation that had spanned the Equator and had apparently emanated from the middle of what is now the Indian Ocean. Biogeographers meanwhile had found similar evidence of widely separated animals that could not possibly have migrated across the waters that now separated them. The obvious explanation at the time was that the intervening ocean had not always been there. Where was that land now? The only reasonable explanation seemed to be that it had sunk.

As the persistence of Atlantis, Lemuria and Mu myths attest, 'sunken lands' tap into something deep in the human psyche, and many theories have been advanced as to why this should be. One has it that, after the last Ice Age, sea levels rose 125 metres in a relatively short time. The sea reclaimed vast areas of coastal land that had been exposed during the great freeze. We know that this happened and we know that humans must have witnessed it. Perhaps this event really did leave deep scars and give rise to ancient legends of drowned land, legends that informed early geological speculations.

On the other hand, if you throw a stone into the sea it sinks. Sinking is what rock does. Land subsides. Things fall into holes. It's the sort of movement that seems *natural* for rocks, acting under the influence of gravity.

But although one could explain many troubling facts by supposing that former land ('land bridges' was the somewhat misleading term) had fallen away to become the bed of the sea (separating things that seem too similar to be so far apart), this did not help to explain Wallace's line, across which very different creatures live in such inexplicably close proximity. Nor did it help much in explaining why the southern continents had all been glaciated at more or less the same time, and on opposite sides of the (present) Equator.

The great British geophysicist Arthur Holmes, an early convert to continental drift, who first suggested convection in the Earth's mantle as a plausible mechanism for it as early as the 1920s, wrote in the 1965 edition of his great book *Principles of Physical Geology*:

The . . . climatic dilemma could only be resolved by realising that the deep-rooted 'common sense' belief in the fixity of the continents relative to each other . . . was now in direct conflict with the evidence of the chief witness – the Earth herself. In other words . . . continental drift had to be taken seriously. But mathematical physicists declared [it] to be impossible and most geologists accepted their verdict, forgetting that their first loyalty was to the Earth and not to books written about the Earth.

To see a thing, first you must believe it to be possible. As it was for the Blanford brothers with their bold interpretation of the Talchir boulder bed, the simple act of believing your eyes is very often an act of considerable mental courage. The same went for the faunal zones and the Wallace line. The simplest explanation, such as William of Occam always urges upon scientists, was that the continents had moved *sideways* across the surface of the Earth. But in the late nineteenth century (and for much of the twentieth) that remained too wild a surmise to be accepted.

Nevertheless, after all this confusion and speculation about a lost continent that had never actually existed, the first *genuine* lost continent to be freed from oblivion by the human mind was emerging into the gaze of a new breed of time traveller. A vanished geography, that had begun its disappearing act 250 million years ago, was backing slowly into the light.

4

LAND OF THE GONDS

The hills are shadows, and they flow
From form to form, and nothing stands;
They melt like mist, the solid lands,
Like clouds they shape themselves and go.

ALFRED TENNYSON, *IN MEMORIAM*

Fixed to number five, at the end of the street nearest to Angel tube station in Islington, north London, is a rectangular green plaque put there by the Geological Society of London, announcing it as the birthplace of Eduard Suess (1831–1914), 'Statesman and Geologist'. Sadly, today almost nobody remembers who Eduard Suess was. But he was recognized in his lifetime as one of the greatest scientists of the nineteenth century; one who, in the course of a long and busy life, manned the barricades in a revolution; brought a new fresh water supply from the Alps to another great European capital, Vienna; and tamed that city's floods. He also wrote a wholly remarkable book which made him the first human being to conceive of a long-vanished giant landmass uniting the southern continents. This land still bears the name he gave it: 'Gondwanaland'.

Suess spent most of his student life in Vienna; but three years after he settled there the city was caught in the liberal revolution that swept Europe in 1848, the momentous year in which Karl Marx published

The Communist Manifesto. Suess might have come from a bourgeois mercantile background but he didn't let it hold him back; and for all his politeness to his English friends, he was no Englishman. He was a young, liberal activist mingling with others who, like him, were soon to take a decisive role in their country's affairs, and who were of an age (and disposition) to man barricades. Suess learnt, in 1848, that the world could change, suddenly and permanently. What is more, sudden revolutions were not only possible: they could do you good. Not that it did him much good at first.

Like many things revolutionary, it started in France. The 1848 Paris revolution, which eventually led to the short-lived Second Republic, caused a run on the Vienna stock market. There was revolution in Austria-Hungary as the rising middle classes demanded change. Reformers roamed Vienna's streets demanding the resignation of Prince Metternich, the widely hated conservative Chancellor, who ruled the Empire in place of the feeble-minded Habsburg Emperor, Ferdinand II (1793–1875). The protesters wanted such things as a free press, freedom of assembly and a national German parliament.

After trying to placate the populace with half measures, Metternich and the Emperor fled to Innsbruck. Barricades, Paris-style, were set up in Vienna's streets – and the young Eduard Suess was on them.

As in other countries, the 1848 revolution in Austria was inconclusive. By August the Imperial family had returned. Sentiment among the ruling elite swung back, hankering after stability. The government camp rallied with the aristocracy and others keen to see the Habsburgs back in power. On 23 October 70,000 troops besieged and bombarded Vienna, against at most 40,000 rebels who included students and academics. They were doomed. No help arrived, and after three days' fighting 2000 of them lay dead. The leaders of the uprising were rounded up and court-martialled. Nineteen were sentenced to death.

Suess who had been sent away for his own safety, now returned to Vienna and continued his studies, yet remained under suspicion, because in December 1851 he and a number of others at the Poly-technikum suspected of allegiance to the Hungarian nationalist leader Lajos Kossuth (1802–94) were arrested, subjected to court martial and imprisoned.

Suess was now in trouble. He might have remained at the Emperor's pleasure for much longer than he did had it not been (if we are to believe some sources) for the intervention of a powerful mentor, Wilhelm von Haidinger (1795–1871), founding director of the Austrian Geological Survey, who used his influence to get his protégé freed in 1852 without indictment.

The great old geological surveys of the world mostly date from this phase of the Industrial Revolution, when governments began to real-ize that everything society needs that cannot be grown has to be found by a geologist. The Austrian Survey was founded in 1849 at the former Mining Museum in Heumarkt, making it one of the oldest in the world. In 1851 the geologists removed to a more prestigious address, Vienna's Rasumovsky Palace. This gigantic pile, on what were then the outskirts of Vienna, had been built, accidentally burned down and then rebuilt, by the former Russian ambassador to the Austrian Court, Andrei Kirillovich Rasumovsky (1752–1836), now chiefly remembered as a patron of Beethoven.

Suess had published his first scientific paper in 1850 (on the min-eral waters of Karlsbad) and another in 1851, the same year that Haidinger commissioned him to map sections across the Dachstein region of the eastern Alps. This work sparked Suess's lifelong inter-est in the structure of mountain ranges and was to lead to one of his most lasting contributions to science. But that lay far in the future. For the time being he needed gainful employment and finally, in 1852, he secured it. It didn't sound like much – clerk in the Imperial

Geological Museum – but Suess was launched. In 1857, still only twenty-six, he completed a spectacular feat of counter-jumping by being made the first 'extraordinary professor' of geology at the University of Vienna. Promoted to full professor ten years later, Suess remained in post for his entire career, retiring in 1901: '88 semesters later', as he was to say in his valedictory lecture.

Man of substance

Suess was no prisoner of the ivory tower. Five years after joining the professoriate he published a pamphlet lambasting the typhus-ridden water that the Austrian capital's citizens were forced to drink, and proposing a dramatic solution. As he later wrote in his *Erinnerungen* (*Memories*): 'The basic principle [is] that drinking water is to be looked for outside settlements.' Suess joined Vienna's City Council in 1862, the year his pamphlet hit the streets. From this position he pushed forward the first *Wiener Hochquellenwasserleitung* – Vienna Mountain Spring Water Pipeline – which eventually solved the city's drinking-water problem in 1873 (and is still used today).

Suess was made an honorary Burgess, Vienna's highest civic honour. Soon afterwards he was chosen as a parliamentary representative and subsequently sat for more than thirty years in the Austrian Parliament, for three of them as leader of the Liberals, raising hackles with his forthright anticlericalism and strident denunciation of political privilege. In many ways he became the Austrian version of Britain's Thomas Henry Huxley: a very public scientist indeed, doughty, rebellious, controversial, yet fully engaged in public works and showered with more honours than he ever accepted.

Perhaps because of his political commitments, Suess's scientific life was not filled with the sort of relentless travelling that geologists usually indulge in. Instead he used his academic resources to survey the world from his study, with his almost unbelievably wide command of literature: from the latest research reports from the far-flung corners of the world's great empires to the *Epic of Gilgamesh*. It was soon recognized across the world that no man alive could match his knowledge of the globe. Between 1885 and 1909 he distilled this unrivalled world-view in a monumental four-volume book called *The Face of the Earth* (*Das Antlitz der Erde*). Even before it was completed the Scottish geologist Sir Archibald Geikie was writing of this 'noble philosophical poem in which the story of the continents and oceans is told by a seer gifted with rare power of insight into the past' and of its 'firm hold of detail combined with singularly vivid powers of generalisation'.

Oddly to modern eyes, Geikie's long tribute to Suess (published in 1905 as the third volume was in preparation) does not mention the book's biggest claim on history, the greatest tribute to those very vivid powers of generalization; for as early as volume one, Suess discovered and named his lost supercontinent.

Scott of Gondwana

In 1913, a year after Captain Robert Falcon Scott had frozen to death returning from their failed attempt to be first to walk to the South Pole, his second in command, Edward Evans, was returning to New Zealand aboard the *Terra Nova* and composing a letter to his secretary and immediate family. Scott's legend had yet to be printed, and Evans's letter, written before the myth of the Great British Hero stifled all criticism of the man, was frank. 'It seems to me extraordinary that in the face of such obstacles *they stuck to all their records and specimens* . . . We dumped ours at the first big check. I must say I considered the safety of my party before the value of the records and extra stores – not eatable.'

To find out why Scott of the Antarctic died lying beside thirty-five pounds of rock, you have to go back to 1905 and a dinner he ate in Manchester in the company of a young lecturer in the University's Department of Botany. That woman, the first female ever to be employed as an academic by the University, was Dr Marie Stopes (1880–1958).

The world knows Marie Stopes today for her later pioneering work on birth control, just as in her day the readers of *Married Love* or *Enduring Passion* assumed that she was a doctor of medicine. In fact, she held a DSc from London University and had already had a distinguished career as a palaeontologist specializing in fossil plants. Much of the research for which that degree was awarded had concerned the evolution of the seed. According to her biographer Keith Briant, Stopes, having met Scott at this dinner dance, quizzed him during the waltzes about his Antarctic ambitions and urged him to take both her and his own wife to the frozen continent. At the end of the dinner, following Scott's persistent polite refusal, she is then said to have urged him, as second best, to find for her the fossils she most wanted.

Seven years later, on 20 March 1912, the exhausted Scott, Henry Bowers and chief scientist Dr Edward Adrian Wilson put up their tent against the blizzards for the last time. Their frozen bodies had to wait until 12 November to be found. The leader of the relief party, Edward Atkinson RN, wrote: 'We recovered all their gear and dug out the sledge with their belongings on it. Amongst these were 35lb of very important geological specimens which had been collected on the moraines of the Beardmore Glacier . . . they had stuck to these up to the very end, even when disaster stared them in the face and they knew that the specimens were so much weight added to what they had to pull.'

However, Scott was not deceived about the importance of the geological specimens. In all, 1919 samples from the 1911–12 Antarctic adventure are now housed at the Natural History Museum in London. Most were collected by expedition geologists Raymond Priestley, Frank Debenham and T. Griffith Taylor, and come from McMurdo Sound and Terra Nova Bay. However, the ones you pick up with most trembling of the hands are those that lay for eight months beside the bodies of Scott, Wilson and Bowers just 12.66 miles south of One Ton Depot. There are samples of coal and fossil plants, and among them is the first find from the Antarctic of *Glossopteris*.

French geologist Adolphe Théodore Brongniart (1801–76) coined the name *Glossopteris* for a fossil leaf in 1828. At the time he thought he was describing part of an extinct fern. *Glossopteris* means 'tongue fern' in Greek, and its leaves are very like those of *Asplenium*, the familiar houseplant Hart's Tongue Fern. However, the *Glossopteris* plant held quite a few surprises. First it turned out to have been a tree which grew to about eight metres; and, despite its fern-like appearance, it produced *seeds*. Such 'seed ferns' became extinct in the Triassic Period, well over 200 million years ago. Today they are studied by a few dedicated palaeobotanists, modern inheritors of Marie Stopes's scientific interest, as they provide fascinating insights into how reproduction by seed evolved. *Glossopteris,* on the other hand, is a name famous among all geologists because it planted, in the mind of Eduard Suess, the first idea of Gondwanaland.

Suess painstakingly pieces together the evidence for his supercontinent. He notes in volume one that India and South Africa have many things in common. Each supported very similar sequences of rocks: 'a mighty series of non-marine deposits which extend from the Permian to the Rhaetic . . . a series of similar terrestrial [fossil] floras flourished in both regions . . .' Madagascar, too, shared this similarity. Finally, near the end of the first volume, Suess utters the sentence: 'We call this mass Gondwana-Land, after the ancient Gondwana

A reconstruction of the *Glossopteris* tree.

flora which is common to all its parts.' And so the first *true* supercontinent was born – *reborn* – in the mind of Man.

Suess was a great coiner of words. Geologists use them all the time
without realizing it. Words like *sima, Tethys, Panthalassa, epeirogen-
esis, syntaxis* and *eustasy*, unfamiliar to most people, have sewn Suess
into the fabric of geological language. Yet even everyday speech bears
him testimony. Every time we talk of 'the biosphere', meaning the
sum total of all living things on Earth, we invoke Eduard Suess. None
of his many terms, however, has caused as much etymological controversy as 'Gondwanaland'.

Suess chose the ancient name meaning 'Kingdom of the Gonds'.
The Gonds, like the Tamils, were Dravidian peoples, and once inhabited much of what is now Madhya Pradesh, in the heart of peninsular
India. Their kingdom's name, Gondwana, had already been attached
by palaeontologists to the typical fossil plants that seemed to link the
now far-flung southern continents, in the term 'Gondwana

Assemblage'. For Suess 'Gondwanaland' meant the land whose rocks yielded this assemblage.

By the time Suess came to write his last volume, his lost super-continent had grown to include: 'South America from the Andes to the east coast between the Orinoco and Cape Corrientes, the Falkland islands, Africa and the southern offshoots of the Great Atlas to the Cape mountains, also Syria, Arabia, Madagascar, the Indian penin-sula, and Ceylon.' Its characteristic fossil plants had now been found in the Permian rocks of South America, South Africa, India and Australia. Once the specimens salvaged from Scott's last tent had made it back to London, Gondwanaland would extend its dominion even further, and include the great frozen continent of Antarctica. This last fact never made it into Suess's book, as it was first revealed in print five years after the last volume of *Das Antlitz* came out, in the same year that Suess died.

Though an obvious point, the main thing to remember about supercontinents is that they have vanished – for the moment. As Suess was the first to find, bringing a lost supercontinent back to life was as nothing to the problem of accounting for its disappearance. Just where did Gondwanaland go? 'Gondwanaland was a continuous con-tinent. Then it broke down, sometimes along extensive rectilinear fractures, into fragments,' Suess wrote. A modern reader, encounter-ing this with a head full of continental drift, might conclude that Suess was not only father of the supercontinent but of continental drift as well. But this would be wrong. In Suess's worldview the Earth was shrinking inexorably as it cooled; his lost continents had sunk, like those lands of myth Atlantis, Lemuria and Mu, beneath the sea.

5

FROM OUT THE AZURE MAIN

Sit down before fact as a little child, be prepared to give up
every conceived notion . . . or you will learn nothing.

THOMAS HENRY HUXLEY

For all the eternity of the ocean, there is nothing timeless about its
shoreline. Over great spans of geological time the ocean may invade
large tracts of the continents and create vast, shallow seas, as it did
when chalk, for example, came to be deposited over so much of the
Earth. Also, ice ages pull water out of the oceans and pile it up on
land, causing the global sea level to fall hundreds of metres and leaving the continental shelves (those fringes of continent temporarily
covered by the sea) fully exposed.

The crust of the Earth itself also goes up and down. When I was
a research student, working on the Baltic island of Gotland, I had the
good fortune to know a local architect named Arne Philip who had
a passion for geology. He would criss-cross the island in his MG con-
vertible, theodolite in the back seat, making surveys of thousands of
the ancient beach ridges that ring the island and all its outlying islets,
recording the high-water marks achieved in the Baltic region thousands of years ago. Arne collected shelves full of data on these 'storm
beaches' and plotted the results on gigantic maps. I do not think, however, that he ever reached any firm conclusions.

The reason is this: when ice ages end, and the heavy ice is removed after tens of thousands of years, the land recovers, just as a cushion does when you get up. The Earth, on its vast timescale at least, is soft to the touch, and this recovery has been happening in the Baltic for 10,000 years. But Arne's problem was that his equation had (at least) two variables. At the end of the Ice Age the sea level rose globally; but locally the land was also rising where it had been covered by thick ice sheets. Gotland's storm-beach heights were therefore a function of two unknown variables – and almost impossible to disentangle.

At least Arne knew what he was dealing with, and understood that the Baltic region is rising and why. Suess did not enjoy the benefit of such well-established explanations. In his day the relative ups and downs of crust and sea over geological time were hotly contested issues, attempted explanations for which came to occupy much of his gigantic book. This plastic quality, this mobility of the rocky crust (the very fact that, given time and sustained pressure, rocks *flow*), eventually proved crucial to moving beyond Suess's contraction theory as an explanation for the break-up of Gondwanaland, and towards building a truly accurate map of the supercontinent he discovered. Yet although geologists now regard as obvious this plastic behaviour of rock over long periods, most non-geologists are still quite surprised to hear how fickle the relationship between the land and sea can be: and not just in Scandinavia.

An island that went away

One day in late 2002 a mysterious volcanic island reappeared in the middle of a busy shipping lane off Sicily. A diplomatic row quickly flared over whose territory this resurgent island would be. Did previous claims still hold? Opinion was hot, strong and diverse. And caught in the middle of it were geologists: the one group that knew most, and cared least for territorial squabbles.

This speck of strategically important potential land has become known to the British as the Graham Bank volcano (and to the Italians as Isola Ferdinandea, the French as L'Isle Julia, and to various others as Nerita, Hotham, Corrao and Sciacca). Professor Antonio Zichichi, a geophysicist at the Ettore Majorano Centre in Sicily, put it simply, and perhaps with a little exasperation, when he told the Belgian newspaper *Le Soir*: 'It goes up and down because the Earth's crust goes up and down and that's that!' But geologists were not always so apparently uninterested in the question of the Earth's crust's ups and downs.

It was Enzo Boschi of Italy's Institute of Geophysics and Volcanology who seems to have rekindled this century's new-found interest in the Graham Bank, in an interview with Reuters, following reports of water disturbance in the area. Dr Boschi had said the volcano might erupt 'in a few weeks or months'. This quote quickly spread through the online media and resulted in a thoughtful feature by Rose George in the UK newspaper the *Independent*. Then the whole thing went quiet for a month, until the Italian Naval League re-awakened the story by demanding that Italy do something to prevent perfidious Albion (or any other foreign nation) from stealing their island again.

The thing about volcanoes is that they erupt from time to time, and as Professor Zichichi knew well, when they do this they tend to swell up as the magma rises (and subsequently down after it erupts). The new activity of 2003 was just the latest phase in the life of the Graham Bank volcano. The previous time it had appeared above the water was in 1831. In the very year that Eduard Suess was born, a small Mediterranean island a few hundred metres across was also coming into the world, just off Italy's toe.

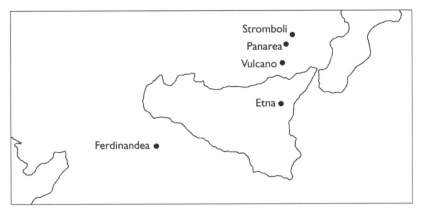

Location of Ferdinandea, or Graham Bank as the British have it.

The British name derives from Admiral Sir James Robert George Graham, First Lord of the Admiralty, who claimed the newly emerged island for Britain by planting a flag on it. This did not stop competing colonial claims by France, Spain and of course the Kingdom of the Two Sicilies. The Italians, whose nation now encompasses the Two Sicilies, still refer to Graham Bank as Ferdinandea, after King Ferdinand II, who ruled the Two Sicilies from 1830 to 1859. The competing colonial nations only gave up their territorial squabble when they realized that the sea had eroded the island away completely while their backs were turned.

Graham Bank sometimes seems to have been put on Earth simply to make fools out of men. As recently as 1987 US warplanes spotted it, believed it to be a Libyan submarine and dropped depths charges on it. Even more recently the two surviving relatives of Ferdinand II commissioned a plaque to be affixed to the then still submerged volcanic reef, claiming it for Italy should it ever rise again. Astute readers will not miss the logical difficulty inherent in the concept of a 'submerged island'. For is not a 'submerged island' just another bit of the seabed? But neither national sentiment nor *fin de race* royals find much use for logic.

The prime mover in the affair with the plaque was one Domenico Macalusa, surgeon, voluntary 'Inspector of Sicilian Cultural Riches' and, as luck would have it, a keen diver. He first became interested in Ferdinandea in February 2000, when news of fresh eruptions first broke. These reports were all couched in terms of the reappearance of a 'lost corner of the British Empire', thus ruffling some Italian plumage. The well-connected Macalusa successfully persuaded Charles and Camilla de Bourbon to commission a 150kg marble tablet, which he and some friends duly installed, twenty metres under the waves, in March 2002. (Mysteriously, it was later smashed into twelve pieces by a person or persons unknown.)

That year Filippo D'Arpa, a journalist on a Sicilian newspaper, published a timely novel about the events of 1831: *L'isola che se ne andò* (*The island that went away*). He summed up the fiasco well when he told the *Independent*: '[Ferdinandea] is a metaphor on the ridiculousness of power. This rock is worth nothing, it's no use as a territorial possession, and yet the French and the Bourbons ... nearly came to war [and] 160 years later, England and Italy are still fighting.'

If only politicians would listen to geologists, perhaps they would learn to curb their enthusiasm. For alongside this grand *opera buffa*, the geology of Graham Bank is really rather boring. Today the volcanic scoria that surrounded the vent in 1831 have been planed off: to thirty metres deep. A column of rock twenty metres in diameter – the lava-choked neck of the old volcano – rises to a dangerous eight metres below sea level (and looks, one must suppose, rather like the conning tower of a submarine to the pilot of a B52). To a geologist it is just alkali olivine Hawaiite basalt, a typical piece of ocean floor. Most of our planet is covered in similar stuff. No wonder geologists were so nonplussed by the public interest in this here-today-gone-tomorrow island.

But in 1831 men of science exhibited no such disdain. Its emergence was greeted as an amazing prodigy, and the most advanced scientific thinkers of the time seized on it with glee. The man who founded the Société Géologique de France, Louis Constant Prévost, wrote a lengthy scientific paper about the island (which he called L'Isle Julia); though by the time it finally appeared in print, in 1835, the island itself had long vanished. Meanwhile in England there was another geologist for whom the movement of the Earth's crust was a source of interest.

Charles Lyell is undoubtedly one of the most influential geologists who ever lived, even outstretching Suess's shadow over subsequent ages. When you read Suess's obituaries you are struck by the way all of them reach for comparisons; for writers with comparable scope, who published great books that will stand for ever as their monument. The one name they all quote is Charles Lyell, author of *Principles of Geology*.

Through his *Principles*, which Darwin took aboard the *Beagle* as his bedtime reading, Lyell had come to dominate the way geologists (especially English geologists) in the nineteenth century thought about Earth history. He preached a strict version of a creed known as uniformitarianism, the concept credited with putting an end to wild speculation about the past and turning geology into a science.

So what is it? The essence of the principle is simple. It says that to understand what the rocks are trying to tell you, you should look around at the causes operating today and find an explanation among them. Geologists constrain their interpretations of the rock record by looking at the way the modern world works. The modern world is the control on geologists' thought experiments.

This is how it works. If you were to see me in the street with a black eye and grazed elbows, you could devise a number of possible scenarios to explain how I got that way. You might conclude I had been abducted by aliens and used as a guinea pig in their experiments. That explanation would not be uniformitarian because, although some pretty unusual things do happen in Stoke Newington, alien abductions are not among them. On the other hand you might surmise that after one glass of Pinot Grigio too many I had missed my footing and measured my length in the gutter. This *would* be a uniformitarian conclusion, because similar events happen almost every day (though not to me, you understand).

As well as urging present-day processes upon geologists, uniformitarianism also has something to say about their intensity. In addition to looking around for modern-day causes, the strict Lyellian assumption is that those processes have also always operated at a comparable rate. Thus deep time becomes paramount. The raindrop falling on the stone can, given enough time, move the mountain. Tiny changes, all but imperceptible to us, can achieve everything geologists might want because time is almost infinitely available. There is therefore no need to appeal to great upheavals or catastrophes; the gradual ups and downs of the Earth's crust, as in the Baltic or the Graham Bank, will be enough.

This view of uniformity is an extreme one, but it was the prevailing view in Suess's time, especially in England. The third (1834) edition of Lyell's *Principles* devoted six pages to Graham Bank. It provided Lyell (who trained as a lawyer, and it showed) with a convincing courtroom argument against his catastrophist opponents. But Suess, who also subscribed to uniformitarian principles, had a different mind: one with mountains in it. Geologists who work among the world's great ranges will tell you they leave an indelible stamp on the imagination. The Alps lay at the root of the Romantic revolution, as artists turned to them for inspiration. Mountains were no longer merely inconvenient obstacles but *meaningful*. Suess had cut his geological teeth mapping the Alps and wrote an early book about them.

By contrast, Lyell hardly mentioned mountain building at all in his *magnum opus*. Today this seems very curious. It is almost as though he thought of mountains as a bit embarrassing, a sort of unsavoury fracas from which an English gentleman should avert his eyes. European geologists like Suess found the Alps much less easy to ignore. They knew in their bones that the Alps had something very important to say about the world and how it worked. Something about the beginning and the end of the world seemed locked up in their tumult.

So geologists are still struggling with two types of change: gradual, repetitive Lyellian ones that go in cycles, and secular – one might even say 'Suessian' – changes: progressive, revolutionary once-and-for-all changes after which there is no going back. The history of the Earth is made of both.

In nature, cyclicity is going around all the time. Our Earth goes around the Sun and we have cycles called seasons. The Moon goes round the Earth and we have cycles called tides. Our planet rotates and we get cycles of day and night. This book is about the greatest cycle of nature: from one supercontinent, through phases of break-up, to the reassembly of a new supercontinent over a period of between 500 and 750 million years.

But there are plenty of examples of Suessian change too. Owing to the friction caused by the tides of the global ocean, the Earth is rotating more slowly today than it did yesterday. The moon's orbit takes it a little further away from us each day. Days are longer now than they were 500 million years ago, which also means there were fewer days in the year back then. The Sun is gradually becoming hotter as it uses up its hydrogen fuel. And despite the delaying tactics of radioactivity, the Earth is indeed very gradually cooling down. Changes like this are one-way-only.

Cycles, however, were the essence of uniformitarianism as presented by Lyell. They allowed nature to repeat herself endlessly to the last syllable of time. What attracted Lyell to cases like the Graham Bank volcano and the ups and downs of the Bay of Naples was that they allowed him to make a subtly different point, namely that even if the rocks do speak of catastrophe, gradualism still dominates time.

The Vesuvius eruption of AD 79 left a lot of geological evidence behind. Catastrophes often give rise to more evidence than the uneventful ages that pass between them. But this does not mean that the past was more violent than today; it just means that the rocks are unrepresentative. Like a scandal sheet called the *Geological Record*, rocks scurrilously report everything lurid and gruesome but leave out the everyday stuff. Lyell's Earth was cyclic, placid even, and there was no progression, wave following on wave.

Suess wasn't having this. His was a uniformitarianism for revolutionaries. For Suess there was more to existence than endless repetition. Not everything that goes around comes around. What happens today *can* make a difference, for ever. Suess rejected the idea that processes going on around us now are the *only* yardstick against which to measure the Earth's massive history.

In reconstructing supercontinents even older than Gondwanaland, lands that existed when the Earth was very different, Earth scientists today are able to envision much stranger things than Lyell's philosophy would ever allow them to dream of, and yet still keep their scientific heads. Suess, who also peered deeply into time, lacked the true Englishman's fondness for the *status quo*. This man, who had stood on mountains and barricades, built aqueducts, tamed rivers and discovered a supercontinent, understood something Lyell did not: things need not always have been the way they are.

Endeth the world . . . (not)

In mid-1960, engineers were carving out the Mont Blanc tunnel, which connects France and Italy, through the roots of the tallest Alpine massif. But on 14 July a small band gathered nearby to witness the End of the World, which was supposed to take place at 2.45pm. As the moment approached, women began wailing. A bugler in lederhosen stood up and delivered an impression of the final trump.

Then, unexpectedly, 2.46 arrived.

The cult leader, Elio Bianca, who before becoming a prophet had worked as a paediatrician with the Milan Electric Company, said afterwards: 'We made a mistake.' The next day the *New York Times* ran a story under the headline 'WORLD FAILS TO END'. You could hardly ask for a more succinct statement of strict Lyellian uniformitarianism.

By contrast, the first people to climb Mont Blanc did so at the behest of a geologist, who was more anxious to know about how the Earth began. Horace Bénédict de Saussure (1740–99) put up two guineas for the first person to find a route to the top of '*la montagne maudite*' ('the accursed mountain') after visiting Chamonix for the first time in 1760. It was twenty-five years before anyone made a claim, but in the end it was chamois hunter and crystal gatherer Jacques Balmat, together with a local physician, Michel-Gabriel Paccard, who became the first humans to stand at the summit, on 8 August 1786.

De Saussure himself gained the summit himself a year later and verified its height as 4785 metres (twenty-five metres short, but enough to put it in the record books). And although he later gave up trying to disentangle the fearsome structural complexity of the Alps, de Saussure summed up a whole tradition of European geology when he wrote: 'It is the study of mountains which above all else can quicken the progress of the theory of the Earth.' Understanding mountains and the processes that build them was to unlock the tectonic enigma of how supercontinents form and break up. Crucially, mountains were soon to demonstrate the impossibility of up-and-down tectonics and foundered continents.

In almost his first geological assignment, Suess had discovered evidence of large *lateral* displacements in the Alps, which seemed to show that massive terrains had been moved sideways for large distances. In doing so he unwittingly planted the seeds of an idea that would unravel not only the structure of mountains but help lead eventually to the idea that continents themselves can move laterally. In later work and his *magnum opus* Suess did not ignore lateral displacement; instead he said it was a side effect of ups and downs. For him the basic force governing all tectonics was shrinkage, which caused large sections of the planet's contracting crust to founder. As he put it, 'The collapse of the Earth is what we are witnessing.'

As the Earth's innards shrank, Suess believed, the crust was put under strain. From time to time parts of it would be forced to subside as the rocky outer shell accommodated to the collapsing planet within. The fragmentation of Gondwanaland, he reasoned, was caused by great subsidences, which left parts of the crust standing high as table-lands (Africa, India and South America) and parts deep below the sea. So the formerly connecting stretches of land in between, for example, Africa and South America, or India and Madagascar, had simply dropped and been lost beneath the waves. Gondwanaland had left fragments behind, but it had not *fragmented*. The lost parts of it were still there, sunk beneath the Indian and Atlantic oceans like the lost continents of myth. Because these foundered areas of new ocean were broadly elliptical, Suess said, they tended to leave behind landmasses with pointed ends: for the same reason that a round pastry cutter leaves you with triangular offcuts on the rolling board.

The idea of a shrinking Earth was a powerful one, because it seemed to flow, with all the inevitability of physical law, from the simple observation that it is hot in mines. The further down in the crust you go, the warmer it gets. Heat is escaping from the Earth's interior. And to nineteenth-century physicists this meant that the Earth was cooling. And if the Earth is cooling, it therefore must be shrinking because that is what happens when things cool. (The idea only finally lost support after the discovery of radioactivity, when scientists realized that, because of the heat generated by radioactive decay, the Earth was not in fact cooling at anything like the rate that had been assumed.)

Although Suess was still alive when Alfred Wegener first proposed continental drift in 1912, he remained committed to fixed continents that occasionally sank below the waves. Yet his explanations of how the Earth's contraction could lead to the very lateral displacements that he himself had noted as a young man were never wholly convincing, perhaps even to him.

Later geologists, Wegener foremost among them, merely tipped their hats at his global observations, synthesis and deductions and, freed from the shackles of contraction theory, explained them away using another mechanism entirely: the notion that continents could move *sideways*. And that is how, rather paradoxically, Suess is numbered among the true precursors of continental drift, despite having remained resolutely 'fixist' all his life.

Everest's missing feet

What finally killed off the age-old idea of sinking continents was the discovery that continents simply *can't* sink: they are, in fact, already floating.

Gravity is a property of matter. Every object exerts a certain gravitational attraction on every other, but the force is so weak that only truly massive objects exert it to a degree that we can measure. Obviously the Earth and other planets exert gravity, but if you are using very sensitive instruments, even the extra mass of mountains could be expected to have an effect.

Mapping can be said to be an act of colonization, and the British Raj was keen to reinforce its dominion by surveying the Empress's possessions with the most modern techniques then available. Mapmakers criss-crossed India using two methods to determine their position, one providing a check on the other. The first of these fixed positions on the ground like a sailor at sea, using the stars, the horizon and a sextant. The other method was triangulation, whereby each point on the ground is fixed relative to another by measuring the intervening distance and taking the compass bearing from each triangulation point to two others. The rest is trigonometry.

When, during the mapping of the Gangetic Plain, south of the Himalayas, these two methods were found to give widely differing results, the mapmakers found themselves in a spot of bother. It came to a head over the difference in latitude between the towns of Kalianpur and Kaliana. These were supposed to be 370 miles apart. But their latitude measurements, determined using the two methods, differed by 550 feet. This did not much please India's Surveyor General, Colonel George Everest.

Astronomical measurement depended on the use of a plumb bob to level the instrument before readings were taken, and Everest had the idea that the extra gravitational attraction of the Himalayas might have been pulling the plumb away from true vertical. The Archdeacon of Calcutta, John Pratt, a Cambridge-educated mathematician, was recruited to examine the conundrum; but his first results singularly failed to make things clearer. When Pratt compensated the astronomical readings for the expected extra gravitational attraction exerted by the mass of mountains *that he could see*, the observed discrepancy turned out to be much smaller than it should have been. The mountains were exerting less of a pull on the plumb bob than they should have done. It was as though they were hollow.

When Pratt continued correcting readings taken in places near to the coast, the reverse was true. The ocean, despite its thick covering of less dense water, seemed to be pulling the plumb bob much *more* than it should have done. Pratt and the mapmakers were on the verge of one of the most fruitful discoveries in all geology. The Archdeacon wrote a paper for the Royal Society.

One of the things that makes science scientific is the fact that reputable journals will not publish anything before receiving the comments of one or more of the expert referees to whom they send every paper. This peer review remains a cornerstone of reliable science. It fell to George Biddell Airy, the Astronomer Royal, to review Pratt's paper for the Royal Society. And it was he who came up with

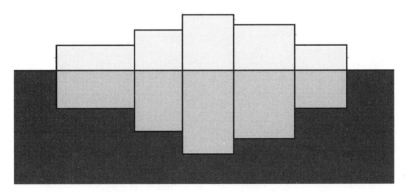

A cartoon showing how, according to the Airy hypothesis of isostasy, like blocks
of wood floating in water, mountains stand high because they are thick.
Their light 'roots' lie immersed in the more dense rock of the mantle, like
the counterbalancing nine tenths of an iceberg below the waterline.

the geologically more correct explanation of these puzzling gravity anomalies. Mountains, Airy said, exert less gravitational pull than they should do because they have roots. Their less dense material extends down into the planet, in whose denser interior they float like icebergs in water. Continental masses, Airy said, stand high above the ocean floor because they are buoyant; in their case, floating in a substrate of denser rock. They stand proud, but only because they have much larger roots below. Mountains are higher than plains for the same reason that big icebergs stand taller than small ones.

The ocean floor, on the other hand, is made of denser rock. To change the analogy from ice to wood, if continents are light, like balsa wood, and stand high in the water, the ocean floor is like mahogany or teak: so dense that it floats, but only just. Hence, despite all that water on top of them, the oceans still exert more gravitational attraction than scientists had expected.

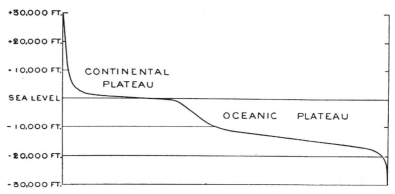

FIGURE 1.—*Generalized Profile, showing relative Areas of the Earth's Surface at different Heights and Depths.*

The original 'hypsographic curve' by Grove Karl Gilbert, showing that the surface of the Earth lies mostly at two levels – average continent and average ocean. Mountains (extreme left) and ocean trenches (extreme right) account for very little of the Earth's total surface area. Gilbert deduced that this graph said something fundamental about the way in which the Earth's crust was organized. Reproduced by permission of the Geological Society of America, Boulder, Colorado © 1893.

Later it was discovered that if you make a graph of the Earth's crustal elevation against the total area lying at that level, on this broad scale (at which small ups and downs can be neglected) the crust has only two basic levels. Continents are almost everywhere a few hundred metres above present sea level, and ocean basins are almost everywhere four to five kilometres below it. The continents have the odd mountain that's very high and the oceans have the odd trench that's very deep. But, basically, nearly all land is at one level and nearly all ocean floor is at another.

This is because ocean crust has its characteristic density and is the same everywhere (basalt), while continental crust is lighter and sits higher. And finally, there's just enough water in the ocean basins to fill them, so nearly all continent is also land and nearly all ocean floor is under several kilometres of water.

This principle is called isostasy, but it is really no more than Archimedes' Principle applied to rocks, which, contrary to all intuition, are all floating. Continents, despite what everyone thought they knew, and despite all the legends and myths, simply cannot sink. True, if you freight the land with thick ice sheets, then the extra mass of ice will gradually cause material underneath slowly to flow away. But when the ice melts, the deep, hot rock will flow back, and the land will rise again.

Although it took time, the idea of isostasy, of the buoyant balance of light and dense rock types, and the knowledge that, given time, the Earth is indeed soft to the touch, was what ultimately paved the way for a true understanding of how supercontinents form and disperse. They do it by moving sideways.

Old world, new world

In 1896, acknowledging the Wollaston Medal, the Geological Society of London's highest honour, the sixty-five-year-old Eduard Suess wrote that it came to him 'at an age when the natural diminution of physical strength confines me to my valley and my home; hammer and belt rest on their peg, and dreams and remembrances alone still carry me along those Alpine wanderings which form the highest charm of our incomparable science, and in the lonely grandeur of which Man feels himself more than ever a child of surrounding Nature'. His last years, devoted to writing his memoirs, were, according to the accounts, peaceful and happy.

Suess's great book was completed nine years after he had retired from the university. His faithful servant, who had continued to bring and take away armfuls of books each day, quite unaware that his master had discovered a continent vaster than any man would ever see, looked at the foot-thick tomes, shook his head and said: 'Is that all you managed to get out of those books I brought you?'

As a boy I inherited my father's partwork encyclopaedia dating from about 1935, entitled *The World of Wonder.* This unwieldy tome contained popular and improving science and engineering articles. Edited by one Charles Ray, with copious but dingy black-and-white pictures and diagrams, it came in regular sections like 'The Romance of Engineering' or 'Wonders of Land and Water', punctuated by illustrated features with headings like 'Inside a great Scotch boiler' and my particular favourite, all about resonance, entitled 'How a small girl can play a big trumpet'.

In year one of its publication, on page 172 there appears a short caption, set beneath three pictures: the Earth with its mountain ranges, a drying apple and an old man's hand. It reads: 'As the Earth gets older, its face wrinkles more and more, just as an apple wrinkles when it becomes dry and shrinks . . . and as the human skin wrinkles when a person becomes aged. Men of science are not agreed as to the cause of the Earth's wrinkling. To some extent it is due to the shrinking of the Earth owing to the loss of interior heat . . .'

Suess's cooling, shrinking Earth appeared to be still alive and well in the 1930s. But change was afoot. Turning to the *World of Wonder*'s page 731 (year two) you find another story, headed 'The Drifting of the Earth's Continents'. Three drawings show the world today, '3,000,000 years ago' and 'as it possibly appeared 200,000,000 years ago': all lands locked together in one outline, one supercontinent, which we call Pangaea, and whose southern lobe bears the legend . . . 'Gondwanaland'.

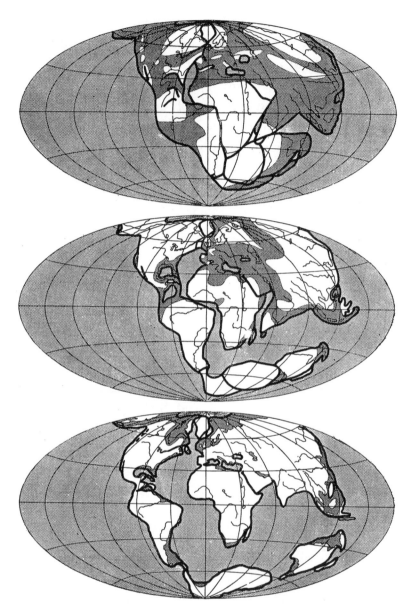

The iconic diagram by Alfred Wegener, showing Pangaea (consisting of Laurasia in the North and Gondwanaland in the South) slowly splitting up to form the continents we recognize today.

PART TWO

EXISTENCE OF LAW

6

WONDERLAND

Geologist finds hole that lured Alice

THE TIMES, 15 SEPTEMBER 1999

Down the rabbit hole

'Down, down, down. Would the fall never come to an end? "I wonder how many miles I've fallen by this time?" she said aloud. "I must be getting somewhere near the centre of the earth. Let me see: that would be four thousand miles down, I think—"'

Lewis Carroll's *Alice's Adventures in Wonderland* (1865) was originally called *Alice's Adventures Underground,* and its opening chapter makes it easy to see why. But the geological connections of this tale may go deeper than Alice's (fairly accurate) reflections on the radius of the globe.

Lewis Carroll was an Oxford University mathematics don, whose real name was Charles Lutwidge Dodgson. He was born in Daresbury, Cheshire, in 1832, the son of a local curate. The family later moved across the Pennines to Yorkshire when Carroll's father became a Canon of Ripon Cathedral. In that cathedral town Charles senior had a friend and colleague named Canon Baynes Badcock. Badcock was the Principal of Ripon College, a Church of England teacher-training school established in 1862. He lived in a house called Ure Lodge, which took its name from the river that flows through this prosperous market town.

The Badcock family included a young daughter by the name of Mary: the model, many believe, for Sir John Tenniel's illustrations of Alice when *Wonderland* came to be published. Lewis Carroll, a prolific pioneering photographer, is said to have passed her picture to the artist. However, amid this Trollopean idyll of rural canons, Oxford dons and genteel pursuits, dark forces were at work.

The district around Ripon is renowned for an alarming geological phenomenon: the sudden appearance of deep, vertical pits, often ten metres or more across, that can swallow trees, gardens, garages and even homes in seconds. These holes are not usually much deeper than twenty metres (though that is deep enough when the remains of your house are at the bottom of it). Worse, though many are dry, some are filled with churning water. As landscape features they are unstable. Left to their own devices, they gradually widen out as their walls collapse, evolving into shallow, conical depressions that may become ponds, marshes or empty hollows. Then they tend to be forgotten.

The fields opposite Canon Badcock's home displayed many examples of this phenomenon, both old and new. In 1837 a particularly big hole developed: one the young Carroll may have seen, either on a visit, or when his family moved to Yorkshire in autumn 1843.

The annals of the Diocese of Ripon record similar collapses down to the present day. And in 1997 local and national newspapers reported that on the 23rd and 24th of April a large crater opened up outside the house of Mrs Jane Sherwood-Britton in Ure Bank Terrace, within the garden of Canon Badcock's former home. The great maw quickly swallowed up two out of a row of four garages, and narrowly missed Mrs Sherwood-Britton's two children, who had been playing beside them only minutes before.

If you visit Ripon you will notice that on the parapet of the Town

Hall is inscribed the pious motto 'EXCEPT YE LORD KEEP YE CITTIE YE WAKEMAN WAKETH IN VAIN'. The annually elected Ripon Wakeman is an ancient office that may date back to Saxon times, though the earliest firm evidence is from 1515. In 1604, with the coming of a new City Charter, the Wakeman was restyled Mayor; but his original job was to keep the peace and act as watchman from the blowing of the Horn, in the Market Place at 9pm, until dawn. (The Hornblower, incidentally, still does his stuff today.)

The Lord may keep the city of Ripon, but to any inhabitant unfortunate enough to suffer from one of its occasional subsidences, the Lord is not much help. English law is distinctly quirky about landslip and subsidence, for which it holds landowners responsible, even though, you might think, such natural phenomena fit the bill perfectly as 'acts of God'. As far as subsidence is concerned, landowners' faith is better placed in geologists than the Almighty.

Nature rarely respects property boundaries, and legal argument at Ure Bank Terrace has since concentrated on demonstrating that no single proprietor can be held responsible for the whole of the hole, two of the four garages that vanished having belonged to two other people, one the modern owner of Canon Badcock's Ure Lodge. And, bringing the story full circle, while the authorities were trying to find that elusive person (who has still to be identified), Ure Lodge itself also succumbed to local geology. Abandoned, empty and crumbling, it was finally transformed by workmen, who took it to pieces and recycled it.

The alarming tendency for bits of Ripon to vanish without warning down deep holes creates many civil engineering problems requiring close cooperation between engineers and geologists. Ripon now boasts a bypass, much of which is elevated above the River Ure on a reinforced embankment. The river flyover had to be designed so that it would remain standing should the ground suddenly vanish from beneath any one of its many well-founded piers. This is comforting to the motorist, because holes like the one that almost swallowed Mrs Sherwood-Britton's house appear on average once every three years, and (on a human timescale) are never going to stop.

In fact, if you tot up the combined volume of the thirty-five subsidence hollows that have formed in the past century and a half, it comes to at least 27,000 cubic metres, according to geologist Dr Tony Cooper. Cooper, an expert on the geology of the area, calculates that under each square kilometre of the Ripon district, at least thirty-six cubic metres of rock are dissolved every year by groundwater.

Tony Cooper recently retired as District Geologist for Yorkshire at the British Geological Survey, where he worked for thirty years. There is a boyish candour and enthusiasm about the way he talks about these alarming subsidences, and an obvious glee when, on his computer, he overlays digitized versions of historic maps one after the other to show how, over the decades, cartographers have mapped 'pits' that in later years became shallow depressions and then boggy hollows. Then there's one map that shows a building on the site, the old pit now filled in and long forgotten. The modern technical sophistication of the Survey's digital resources is a far cry, however, from the hammer, compass-clinometer, pencil and field-slip world of geology in the 1970s.

'When I first joined the Survey I started mapping in Yorkshire,' Cooper told me, 'near York, then Wetherby, Knaresborough, Ripon and Bedale. I started to come across holes in the ground that I couldn't explain. The first were south of Knaresborough. On the floodplain of the river, in a dead straight line, I found four craters. They were fairly sharp-sided, but they had no spoil heaps, so they didn't look like someone had dug them. As I worked north towards Ripon I came across more, and started delving into the old literature. By 1982 I had sorted out the story, and come across quite a lot of other historical material' –

including one Revd Tute and the paper 'On certain natural pits in the neighbourhood of Ripon' that he presented to the seventy-third meeting of the Geological and Polytechnic Society of the West Riding of Yorkshire in the Wakefield Church Institute one April night in 1869.

From the Revd Tute's first steps towards a rational understanding of Ripon's subsidence problems comes our modern knowledge that Ripon's bedrock conceals layers of calcium sulphate – the soluble mineral known as gypsum (raw material for plaster) – in two thick strata.

Tony Cooper saw that prominent joints in the rock – natural partings and lines of weakness – were dictating the sites where dissolution of this gypsum was having the greatest effect. Rocks rarely have just one set of joints; they tend to occur in what are called conjugate sets, rather like the intersecting cardboard sheets in a case of wine. Where one set of joints intersects another, Cooper realized, dissolution was doubly enhanced. That was where the biggest caves formed, so giving rise to the regular, lattice-like pattern of sinkholes dotted about the landscape.

But Cooper was also responsible for another discovery, the link between Ripon's unique geology and the world's most famous children's book. Lewis Carroll's connection with Ripon was already well known: some of his characters were allegedly inspired by carvings on misericords in the Cathedral's choir stalls. If you inspect the misericord under the Mayor's seat (north side, east end) you will see a gryphon chasing two rabbits, an allegory of evil preying on the weak. One rabbit has been caught, but one has dived into its hole – you can see his little scut just disappearing into the Earth. Gradually, as Cooper pored over the old geological literature, he began to notice a recurring coincidence of collapses, clergymen and Lewis Carroll.

By 1997 Cooper's notion about the link between Carroll's imaginative landscape and the landscapes the writer had known as a boy was growing stronger. He began reading up on the author's life and while doing so he also noted an earlier, possibly significant geological association, with the village of Croft-on-Tees, south of Darlington, where the Dodgsons moved when Carroll was eleven years old. Barely a mile from where they lived are some mysterious (reputedly bottomless) ponds called Hell Kettles, one of which was formed by a dramatic subsidence in 1179. To this day, Hell Kettles are filled with water that rarely freezes, visibly steaming in the winter, and giving off sulphurous smells. As at Ripon, the subsidence was caused by gypsum dissolving underground. Such a local landmark and its diabolical legend could not have failed to come to the notice of the inquisitive young writer.

Cooper, who like Carroll is an enthusiastic amateur photographer, carries his enthusiasm for geology and landscape into many other fields, and this excursion into literary research was for him not that unusual. But his theory lacked an outlet until 1999, when the annual UK science festival organized by the British Association for the Advancement of Science rolled into Cooper's *alma mater*, Sheffield University.

The tide of ideas that had been rising in Cooper's mind for the best part of a decade had suddenly found an outlet. Sheffield University geologist Dr Mike Romano was charged with organizing the geology sessions and contacted him to speak about the recent Ripon subsidence. Cooper decided to enliven his story about gypsum dissolution by linking it to something everybody knew about: *Alice in Wonderland*. The rest is a small part of media history.

In fact, everything about Ripon's civil engineering problems is history, except the modern technology to deal with them. The collapses happen because groundwater dissolves gypsum lying beneath the whole region. As caverns open up within it deep underground, their roofs begin to spall and stope their way slowly upward to the ground surface. If you have ever removed a fireplace and then watched in horror as the bricks in the chimney-breast above fall out on to the floor, course by course, you have seen this process at work.

But how did the gypsum get there in the first place? To find that out we need to treat Alice's rabbit hole as a time tunnel to Wegener's lost supercontinent, just as it is beginning to break up.

In the mind a map

Leaving behind Field View, Mrs Sherwood-Britton's old house, you return to the main road at Ripon's medieval North Bridge. Pausing to admire the piers of the new A61 Ripon bypass, with their invisible big feet, you turn through the city centre and head out on the B6265, due west, towards the Pennines: the backbone of England. Passing Royal Studeley, its deer park and Fountains Abbey World Heritage Site on your left, you take the turning to High Grantley and make your way up the rich, gently climbing farmland to High Skelding and Dallow Moor.

Quite suddenly, after the rich farms, their fertile fields and ancient hedgerows dotted with oak trees, the landscape begins to change. Drystone walls replace the hedges, and then themselves vanish as you pass on to the open moor underlain by Millstone Grit, Carboniferous-age rocks that were already deposited, hardened and folded into mountains, as the supercontinent of Pangaea assembled. They are humbler now, of course, after 250 million years of erosion; but as you walk up the heathery slopes to stand on a rough lump of Grit or in the lee of a sheep shelter and look down across the lowlands to Ripon, you are climbing the exhumed topography of Pangaea.

Close your eyes. You hear the wind over the heather, gorse bushes and hawthorns. Larks are rising. Mournful, far-off cries of birds of prey drift across the hillside. It is time to follow the White Rabbit, and imagine his pocket watch racing backwards 250 million years.

You are now just two million years or so into what geologists have named (after the Chinese locality where its deposits are best displayed) the Wuchiapingian Age of the Late Permian. You, and those barren grey mountains of Millstone Grit behind you, whose roots will one day lie exposed to the boots of fell walkers, now sit about twenty degrees north of the Equator; the same latitude as Port Sudan, Timbuktu and Santiago de Cuba today. Although the Sun is slightly less luminous, it is searingly hot, perhaps in the upper forties. From the bare rock at your feet, you take in the slopes of brown, varnished boulders stretching to the plains, a sea of brownish dunes, up to fifty metres tall. Not everywhere down there is covered in sand; in some places rock desert pokes through, and in others, more low-lying, you can see greyish white: the salty fine mud of a desert playa. Towards the horizon its dirty white merges into the mirage. Towering greenish dust devils suck at the dry mud and, driven by the merciless Sun, carry it high into a sky where no bird has ever flown.

Behind you the Pennine range separates you from similar lowlands to the west, but you would have to cross another 4000 kilometres of mountain and desert – nearly all of the future North American continent – before you would come to a coast. But what a coast! This longitude-paralleling shore stretches almost from pole to pole.

Eastward you could travel through all of Europe, across the newly formed Ural Mountains, and across much of modern Siberia beyond, before you would meet sea again. To the south (following the 'N' on your compass because this is a time of magnetic reversal) you could walk all the way to the South Pole. This period, when the north magnetic sits at the south geographic pole, will last for another two million years before flipping back again over the geological instant of a few thousand years. But by that time the eastward view below you (over what will one day be the Vale of York) will have changed beyond recognition.

You feel as though marooned on a raft in mid-ocean, lost in the remotest heart of this seemingly endless expanse of parched land. But though dry, it is not completely lifeless. In the folds of the higher ground, or nestling at the foot of rocks where water comes nearer to the surface or seasonal rainwater collects perhaps once a decade, shrubby horsetail-like plants grow in clumps, relying on their tough, deep rhizomes to find what water they can. Another shrub, *Peltaspermum*, and in rare places some conifers related to pines and the gingko, provide sparse shade. The living is about to get a bit easier for some of these survivors. Still others will be overwhelmed.

Since the beginning of the Permian Period forty million years before, the supercontinent on which you are sitting has slowly drifted about fifteen degrees north. The time of the greatest dying in the history of life, the most severe mass extinction in the geological record, is almost upon the Earth. In the next few million years 90 per cent of all species now living will become extinct. After Pangaea, nothing will ever be the same again.

You might have noticed (had you really been trekking up these same gritstone slopes in the Late Permian) that you had become a little more out of breath than usual, as though you had suddenly become a lot less fit. That is because the air is thin. The atmosphere of the modern world contains about 21 per cent oxygen. About 300 million years ago, before the Permian began, with the Carboniferous coal forests pumping out oxygen from their rampant photosynthesis, it had been higher still, at about 30 per cent. But the atmosphere's oxygen has been tailing off as the great coal forests dwindled. In the late Permian it stands at about 16 per cent. Breathing at sea level has become like breathing at 3000 metres in the modern world; but by the time global oxygen levels bottom out at 12 per cent, as in the next few million years they will, living at sea level will be like living at almost 6000 metres today – higher than any permanent human settlement in the modern world. Little wonder, then, that the four-legged land animals stalking the Permian landscape were having a bad time. Of the forty-eight families of such beasts recognized among the rare fossils found in land sediments, thirty-six had died out by the end of the Permian. Lack of oxygen was one factor among many that drove them to their doom.

The whole globe has also been slowly, fitfully getting warmer. Carbon dioxide and other greenhouse gases have been building up in the air, making life even more difficult. As temperature rises, animals' metabolic rates also rise, further increasing their need for oxygen. And although at this time it may not quite have begun as you sit on the Pennine slopes, far over those distant Ural mountains, in what will one day be northern Siberia, the Earth is about to split asunder in a catastrophe surpassing any biblical horror.

Over a period lasting as little (to a geologist) as 500,000 years, and almost exactly coincident with the disappearance of most living species 250 million years ago, massive eruptions will spew between two and three million cubic kilometres of lava on to the Earth's surface in our planet's biggest-ever series of volcanic eruptions. Today those lavas, known as the Siberian Traps, cover 350,000 square kilometres and are nearly four kilometres thick in places. Significantly for life on Earth, along with that molten rock also came (according to one estimate) 10,000 billion tonnes of carbon, most of which ended up in the atmosphere.

Carbon dioxide will not be the only gas evolved. Sulphur dioxide will also pollute the air and acid rain will fall everywhere. The sea's plankton, bottom link of the oceanic food chain, will be decimated. Rising temperatures will also affect the sea bed. Billions of tonnes of methane, trapped in cold, shelf-edge sediments, will suddenly become unstable and rise catastrophically to the surface, liberating even more carbon into the air than the Siberian volcanoes. Methane is the most powerful greenhouse gas of all, and thus the spiral of environmental breakdown will career towards the greatest extinction in Earth history.

Although there is no ice at the North Pole (and, like today, no land either), the South Pole has been located within Pangaea's southern half, Gondwanaland, for millions of years. There a massive icecap has eroded huge amounts of ancient rock and pushed it out over much of what is today India, South America, Antarctica and Australia; but it has now dwindled almost to nothing. Mud and boulders the size of men have been dumped everywhere as the ice fell back, like the abandoned weapons of a retreating army, for the Blanford brothers to find at Talchir. Water, locked in the ice sheets for millions of years, has poured into the global ocean, Panthalassa, which has begun a fitful but inexorable rise.

The formation of Pangaea itself, coupled with the removal of carbon dioxide from the air by land plants, triggered the great Gondwana ice age. Although this lasted for many millions of years and the centres of glaciation migrated as Gondwanaland shifted relative to the South Pole, the decisive moment came when the northern continental mass joined Gondwanaland at the close of the Devonian Period 355 million years ago. This closed off the equatorial current that helped distribute heat about the globe and Pangaea was born in an 'icehouse' world.

But at the end of the Permian all this is changing. The Earth system is flipping from icehouse to greenhouse mode. The increased volcanic activity associated with the break-up of Pangaea, and the eruption of the Siberian Traps, will both pump greenhouse gases into the atmosphere. The melting of the glaciers is already increasing the Earth's absorption of the Sun's heat. As the submarine spreading ridges become more active, they are swelling up and pushing the oceans, already full of glacial meltwater, on to the continental shelves. This in turn is further raising the heat-absorbing capacity of the planet because water absorbs and retains heat better than land.

The Earth's climate will remain in its new greenhouse state for much of the next 220 million years, until quite close to our own time, as life gradually recovers from the end-Permian disaster, the dinosaurs come and go, birds fill the sky and the supercontinent of Pangaea slowly dissolves, fragmenting into today's map.

As you sit on the proto-Pennines, something of those tectonic processes tearing Pangaea apart is at work below you, under the parched desert lowlands to the east. The seemingly endless plain is feeling the tension as Pangaea's northern half, Laurasia, begins to unzip: a process that will eventually form the Atlantic Ocean. The floor of the plain is subsiding, falling like a keystone into a widening rift, and desert sediment is tumbling in, piling up into layers of porous rock that will one day act as reservoirs for North Sea oil and gas. As the supercontinent rends itself, the tension extends northwards along a relatively narrow belt between the ancient rocks of Greenland and Norway, now lying cheek by jowl.

There in the distant north, perhaps a week or so before your visit, the inevitable has finally happened. Thanks to that subsidence (despite the inrush of sediment that is doing its best to fill the vacuum), the desert plain over which you gaze lies up to 250 metres below sea level: an ancient Death Valley on a vast scale, stretching from here far into eastern Europe. That entire desert basin is about to become sea, and the whole flooding process will take just a few months. These two events, the flooding of the North Sea basin and the eruptions of the Siberian Traps, took place at about the same time in Earth history, and show how misleading the strict uniformitarianism of Charles Lyell can be. Our knowledge of what is normal behaviour for the Earth is extremely limited. Human beings have not existed on the Earth long enough to have witnessed the eruption of a Large Igneous Province. Nor has our species ever seen a major inundation like that about to unfold at your feet; though neither event is that uncommon in the long history of the Earth.

A wild surmise

Take an example from more recent geological time. The modern Mediterranean would not exist if it were not connected to the global ocean via the Strait of Gibraltar, because not enough water flows off the land that surrounds it to outstrip the process of evaporation. But the Strait is shallow, and about six million years ago, during the Miocene Period, global sea levels fell so much that the connection was broken. The Mediterranean dried out, leaving a vast desert basin that was only retaken by the sea 900,000 years later (still millions of years before modern humans were around). If you ever wondered why there are so many deep gorges in the South of France, this is the reason. Starting from their new bases (the bottom of the desiccated Mediterranean), these rivers eroded back rapidly along their courses, cutting deep, slot-like canyons like cheese wires slicing through a slab of Cantal.

Long before this became fact, H. G. Wells was studying geology at London University under his uninspiring teacher John Wesley Judd ('washing his hands in invisible water as he talked'). The science-fiction pioneer learnt about geological speculations that the Mediterranean had once been dry, which were then being bandied around as explanations for some odd distributions of plants and those mysteriously deep gorges. In 1921 Wells incorporated the (then) unsubstantiated theory in a tale he published in the April issue of *Storyteller* magazine about an encounter between modern humans and Neanderthals. It was called *The Grisly Folk*.

'It was in those days before the ocean waters broke into the Mediterranean that the swallows and a multitude of other birds acquired the habit of coming north, a habit that nowadays impels them to brave the passage of the perilous seas that flow over and hide the lost secrets of the ancient Mediterranean valleys.'

Now, Wells's suggested timing for the Mediterranean's big flood was awry by several million years, and as a theory the 'dry Med' remained unproven until 1970. On 28 August the Deep Sea Drilling Project research vessel *Glomar Challenger* was poking a hole in the western Mediterranean floor, south of the Balearic Islands. They were drilling in almost 3000 metres of water at the time, so the geologists on board were greatly surprised when they picked, from between the teeth of their drill bit, chunks not of gypsum exactly but of the anhydrous calcium sulphate mineral anhydrite. You find anhydrite only where there has been evaporation, and you certainly don't find it on the bottom of the ocean – unless of course the ocean had once dried up and then been reclaimed by a catastrophic flood.

Something very like that flood is now unfolding below you as you perch on the Permian Pennines. Geologists continue to debate the relative role of local subsidence versus global sea-level rise, but the evidence in the rocks for the *speed* of the inundation is dramatic and unequivocal.

What first signs of the advancing tide might you sense from your vantage point? Some tang on the prevailing north wind of this desert basin perhaps? A smell like that of rain on parched city streets? Distant thunder as an unseasonal line of thunderheads advances, formed as water vapour convects violently off the desert and climbs miles high into the atmosphere? Or perhaps a sudden increase in the abundance of small reptiles, fleeing before the advancing menace, planting their three-toed tracks in the sands?

Such events as these would leave few or no conclusive traces; but still we can tell the flood was sudden. The evidence lies in the remains of those sand dunes, which today survive as some of the topmost sandstone gas reservoirs of the southern North Sea. The dunes that built those sandstones were huge, some probably over fifty metres tall. You can see them clearly from where you sit. But dunes are just sand: loose, weak, unconsolidated. Think of a sandcastle on the sea's edge; it doesn't take many feeble ripples to plane it flat. But these dunes were not planed flat.

Two hundred and fifty million years later, if you visit quarries where these fossil sand dunes are now exposed in section, you can clearly see how the first bed deposited on top of them by the new sea drapes over those ancient sand hills' original shape. Moreover, the parts of the dunes closest to the interface with water-laid sediments – where the sand could be expected to have been reworked into beaches as the transgression gathered pace – lack any of the features of beach sands (typical bedding, or the shells that should be found if the beach had existed for any length of time). There are not even any fossil burrows. The inescapable conclusion is that the desert became the bottom of the sea far too quickly for normal shallow water features to be established – or even for the waves to plane the dunes flat.

By the time the great flood was over, the new sea was nowhere more than 250 metres deep. That means that the amount of water needed to fill the entire basin (stretching all the way from Yorkshire to Poland and Russia) would be about 110,000 cubic kilometres. All this would have passed through one narrow northern channel; so the greatest uncertainty in the equation comes in estimating how wide and deep that channel was. If it were ten kilometres wide at its narrowest and allowed a flow twenty metres deep travelling at about three metres per second (a rather conservative estimate), the water could have rushed in at the rate of about fifty cubic kilometres per day. The whole process of filling the new sea would have taken about six years.

Six years seems long enough, but the rate at which the process was completed is not important when we are confronted by rocks in outcrop. Face to face with the record of events at a single locality, the question is: how fast could *any single dune* have been covered by water? Even using our conservative estimate of influx rate, it would seem that in what is now Yorkshire, the sea would have risen by some tens of centimetres a day – enough to bury a fifty-metre-high dune in about eight months. One season and the dune sea below you would have become the sea bed, its once sunstruck curves draped in black mud.

The suddenness of the inundation also explains other features characteristic of the dune sandstones underneath the first marine sediment. Dunes have a characteristic internal structure, formed as sand grains are blown over the crests to cascade down the lee slope as the dune migrates downwind. This creates a large-scale form of 'crossbedding', measurable in metres; the fossil dune surfaces forming great rococo festoons and swags.

The odd thing about these particular dunes is how many of them now appear to lack this characteristic bedform, especially at their centres. Here the laminae of sand are often either contorted and chaotic or have vanished completely. For a long time geologists were at a loss to explain this; but the emerging tale of the dunes' sudden inundation provided an explanation. As the dunes were buried, large pockets of air became trapped at their hearts. Eventually, as the water got deeper, this trapped gas would eventually overcome the strength of the sediment confining it and be released suddenly, disrupting all the original bedding of the sand as it escaped.

So, as you watch the advancing tide from your vantage point on the slopes of the Pennine mountains, and see the dune tops slowly vanish beneath a scummy, turbid tide of thick, slimy, bitter water, you will be rewarded from time to time by the sight and sound of sudden bubbling.

Zechstein

Because at least part of this transgression of the sea was caused by the global rise of sea levels, this story was repeated all over the edges of the fragmenting supercontinent. But this particular example, which geologists call the Zechstein Sea, was (like many of the others) not stable. Like the modern Mediterranean, it could not exist for long without being connected to the global ocean. But because the Zechstein was a shelf sea underlain by continental crust, it was much shallower than the Mediterranean, which is a true ocean floored by dense ocean crust that sits low on the Earth's surface. This made the Zechstein especially vulnerable to drying.

This cannot have been a simple 'on-off' process. Zechstein sediments, now buried deep below the bed of the North Sea, are hundreds of metres thick. To make just thirty centimetres of evaporite (as minerals produced this way, including anhydrite and gypsum, are known), you need to drive off five hundred metres of seawater: twice the depth of the Zechstein Sea at its deepest. Clearly, fresh supplies of ocean water had to be entering the sea continuously, over long periods, and evaporating under the intense heat of the Permian desert.

And this, of course, is where Ripon's troublesome soluble gypsum comes from. The Zechstein Sea may have dried up almost completely as many as five times in its relatively short lifespan of barely ten million years. In doing so, it left behind regular cycles of chemical deposits, each series beginning with the most insoluble minerals (which precipitate first) and ending with those that crystallize only when there is hardly any water left to be dissolved in. So the first minerals to appear are limestone (calcium carbonate, which precipitates readily in warm, saturated water and may do so in your kettle) and dolomite, an impure limestone made of a chemical mixture of calcium and magnesium carbonate.

The first such rock to be deposited, known generally as the

Magnesian Limestone, was the very rock chosen to build the grand new Palace of Westminster, home of the British Parliament, which was rising as the Blanford brothers were leaving for India in 1856. Called Anstone, it came from quarries near Worksop, and proved a disaster in the metropolis's acid rain. Alas, despite its workability and lovely biscuit colour, the Mother of Parliaments' new home soon began dissolving before its builders' eyes, giving rise to a lot of amusing but chemically suspect jokes about why the geologists advising the Parliamentary commission had suggested building the Palace of Westminster out of laxative (Epsom Salt is magnesium *sulphate*).

The new sea brought some relief to the barren heart of northern Pangaea, and it is likely that around its edges the land grew green, or at least greener than it had been. But the supercontinent that enclosed it turned the Earth into a very different world from ours.

Immortal cells

The Earth's climate is largely controlled by a set of fairly simple physical constants, but as scientists are increasingly finding, the combination of simple things can have results of almost unpredictable complexity.

As it orbits the Sun, at a distance of ninety-four million miles, the Earth receives a certain amount of radiation from it, known as insolation. The Sun's output has been increasing with time, and over hundreds of millions of years this small increase – brought about by the gradual exhaustion of its primary fuel, hydrogen – is significant enough to have to be taken into account. It is one of those secular changes of which Sir Charles Lyell would not have approved.

This radiation hits the Earth and warms it up, and the atmosphere of the Earth keeps the heat in by the well-known 'greenhouse effect', and moves it around. By and large, the average energy received at the top of

Earth's atmosphere is a fairly constant 343-watts per square metre: a bit more than three lightbulbs' worth. But the complex interaction of axial tilt and other superimposed cycles made the distribution of heat over the surface of the planet a very complex thing to model.

Seasons, the most obvious climate changes of which we are aware, are caused by the tilt of the Earth's axis relative to the Sun, which currently stands at about 23.4 degrees from the 'vertical' (defined as the right angle to the plane of the Earth's orbit around the Sun, called the ecliptic). Thus, as the Earth revolves around the Sun, for half the year the Northern Hemisphere is tilted towards it, while for the other six months it's the turn of the Southern Hemisphere. This tilting effectively concentrates the Sun's heat first in one hemisphere and then in another, just like leaning towards the fire to warm your face (Northern Hemisphere Summer) and then walking around to the other side, so that you face away from the fire and the heat warms your bottom instead (Southern Hemisphere Summer). At the Equator, of course, this axial tilt has little effect and seasonality is less noticeable.

But there's much more to it than that. If you have ever watched a spinning top that isn't moving perfectly, or the behaviour of a double pendulum, you will have a feeling for the complex way in which harmonic systems behave. There are also eccentricities in the system to consider, and cycles that affect the degrees of eccentricity. Several such long-period cycles affect the orbit of the Earth around the Sun, and these in turn change the climate because they affect the amount of insolation: how much heat hits a unit area of Earth in any one place. How all these cycles interact, sometimes reinforcing one another, sometimes cancelling one another out, creates a highly complex system that means that the Earth's climate is never constant.

Take the Earth's elliptical orbit. The Sun does not sit at the centre of the ellipse, so the Earth–Sun distance that every schoolboy thinks he knows is actually only an average. However, this ellipticity varies (over a period of 98,500 years) from very elliptical to almost circular. At its most elliptical, the extra distance from the Sun can cut the amount of insolation by as much as 30 per cent from when the Earth is closest. This cycle has almost no effect at all on the *total* amount of heat received by the Earth per year, because it all averages out. However, it does increase 'seasonality' (the contrast between the seasons) in one hemisphere, while reducing it in the other.

The inclination of the Earth's axis to the ecliptic varies (over a timespan of 41,000 years), between extreme values of 21.39 (nearest to 'vertical') and 24.36 degrees (most inclined). This cycle also affects the length of the dark polar winters, and has a marked effect on climate in high latitudes. In addition, the Earth's axis of spin describes a circle (over a period of 21,700 years). You can make a spinning top do this very easily by giving it a nudge. This is called precession.

If you combine two of these factors – the 98,500-year cycle in orbital ellipticity with the 21,700-year cycle in the Earth's axial tilt (precession) – you generate a harmonic interference between the two cycles: they produce another cycle. At one extreme the Earth will come closest to the Sun during the Southern Hemisphere Summer; and at the other it will come closest during the Northern Hemisphere Summer. At these extreme points in the cycle, the additive effect of axial tilt and proximity to the Sun makes the summer more intense (and, six months later, the winters deeper, occurring as they will, when the Earth is at its farthest from the Sun). Intense summer conditions will increase the heating of land areas (land heats up and cools down much more quickly than the more even-tempered ocean), with striking effects on rainfall, as we shall see. The overall effect creates cycles of seasonality. But by the same token, when the seasons are at their most contrasted in one hemisphere, they will be at their least contrasted in the other.

Orbital climate-forcing effects were first described by Scots geologist James Croll (1821–90) and later developed by Serb mathematician Milutin Milankovich (1879–1958), and for this reason they are known collectively as Croll–Milankovich cycles. But the climate is not all about angles of tilt and rays per square metre. The Earth's fluid shells – the air and water – are what make it completely different from any other space rock struck by starlight. Earth's atmosphere and the hydrosphere absorb and transport the Sun's heat around, creating an equable average temperature at surface (currently about 25 degrees Celsius). In the oceans this circulation is achieved by a set of interlocking convection-driven cells called gyres; there's one in the North Atlantic and one in the South Atlantic, for example. But they are not discrete: they mesh like cogs in a gearbox, shunting water (and heat) from one gyre to another. In fact, the ocean basins are connected by a three-dimensional 'global conveyor', as it has become known, refreshing and warming bottom waters, creating fertile upwellings of cold, mineral-rich waters elsewhere, and preventing stratification: the tendency of warm water to float on cold, light on dense. This keeps the whole ocean system oxygenated and healthy.

Oceanic convection cells are very much dependent on the shape of the ocean basins – and hence on the distribution of continents. But if you want stability, look to the atmosphere. Here three huge, sausage-like convection cells sit around each hemisphere like the folds of rubber flesh surrounding M. Michelin. They are invisible of course, though the cloud patterns give them away – if you know what to look for. They have existed for billions of years and continue their convection more or less irrespective of what the orbit is doing, or where the continents happen to lie on the shifting surface of the globe. Behind the fickle airs there is a dynamic stability that has easily outlasted the transient continents.

These great convection tubes create the major climatic zones of the

Earth, which like them lie in belts parallel to the Equator. The cells exist as the stable answer to the need to dissipate heat from where it is most plentiful – at the Equator – to the poles. At the Earth's waist-line, hot air rises, creating more or less permanent low pressure and rain as moisture condenses. The rising air hits the upper edge of the atmosphere and splits in two, some going south, some north. We shall follow the northern limb.

This air travels north high up at the top of the atmosphere until it meets more – circulating in the next cell – coming in the opposite direction. The two currents collide and sink back to Earth again. This falling air is dry and creates permanent high pressure. Where it hits land it produces desert conditions everywhere on land except near coasts, where some moisture can blow a little way inland. Thus on either side of the wet equatorial region you find bands of deserts. They stand out well on those 'where is the plane?' simulations pro-vided on long-haul flights.

On hitting the Earth, the air splits again. Some goes back south, to pick up moisture and rise again at the Equator. The rest travels north along the Earth's surface and does not rise again until it meets cold air travelling Equatorwards from the pole. The two then meet and rise, creating another line of low-pressure systems, and rain. Over the poles, in the final or Polar cell, cold dry air sinks, creating high pres-sure with (usually) relatively low evaporation – the dry arctic air of the tundra.

Turning in the widening gyre

A complication, introduced by the Earth's rotation, is the Coriolis effect, named after French mathematician Gustave-Gaspard de Coriolis (1792–1843), who worked out the mathematics governing it. This is the apparent force, acting on all objects moving on the Earth's rotating surface, that tends to deflect them to the right in the Northern Hemisphere and to the left in the Southern. This is why weather systems (and, allegedly, water disappearing down plugholes) rotate clockwise north of the Equator and anticlockwise south of it. On air moving in the cells, it acts to change the simple circular, 'up-across-down' convections I have just hinted at into helical ones.

So in fact the winds within the cells spiral around inside them, like the rifling inside a gun barrel. And because these helical convection currents are wound in opposite directions either side of the Equator (coiling to the right in the north and left in the south thanks to the Coriolis effect), they give rise at surface to the famously reliable trade winds, beloved of sailors. The trade winds just north of the Equator blow from the north-east because the Coriolis effect deflects winds travelling south (completing their return leg to the Equator) to the right (i.e., the west). Contrariwise, below the Equator, the south-east trades blow from that quarter because these winds are deflected to the left (the west again) as they travel north.

At higher latitudes than the tropics, the surface winds of the second great cell blow from the south-west in the Northern Hemisphere (bringing Britain its rain from the Atlantic) because those convection currents become deflected to the right, veering westerly. Above the southern tropics, winds that would be blowing back towards the South Pole (i.e., northerlies) are deflected to the left, backing westerly.

What does all this mean for reconstructing vanished supercontinents? To some extent, no matter where the continents lie, the prevailing winds between the Equator and the tropics, and between the tropics and the pole, have always blown, and will always blow, in pretty much the same direction. These winds will be wet in the same places, and dry in the same places. Falling air will create high pressure; rising air will create lows. It's simple – it's physics.

The way in which the atmosphere then interacts with them creates different environments, which the geologist can diagnose by looking at fossils, and the rocks that contain them, and by comparing this evidence with organisms and sedimentary environments around us today.

But then the distribution of land and sea comes into play, and snarls up this simple convecting system. Think of how, joining Laurasia, the northern landmass to Gondwanaland cut off the equatorial currents and plunged Gondwana into a deep ice age. The distribution of continents clearly affects the way the oceans' gyres work, and in much more unpredictable ways than the unchanging and imperturbable Polar, Ferrel and Hadley Cells of the atmosphere. Moreover, the monsoon is entirely dependent on the distribution of land and sea, the heat differential between them, and seasonal temperature differences across the Equator. These elements prevent the atmospheric circulations from perfectly overlaying an unchanging pattern of climatic stripes upon the shifting continents.

Megamonsoon

Children's encyclopaedias never fail to include an explanation of onshore and offshore breezes; the former created during the day when the sun heats the land, and the latter at night when the land cools off quickly, leaving the sea warmer. Monsoons are in a way similar, but writ large, operating at continental scales over annual rather than daily cycles, and large enough to disrupt entire climate zones.

The word 'monsoon' comes from the Arabic *mausin*, meaning 'season', and refers to winds that change from one part of the year to another. However, the common usage of 'monsoon' is for heavy rain associated with the summer monsoon winds of Asia, blowing ashore off the Indian Ocean.

In India, for example, monsoon rains arrive in early summer. The winds blow onshore from the south-west, though these are actually the *south-east* trade winds of the subequatorial Hadley Cell, being pulled off course by hot air rising over the baking heart of India. The Himalayas and Tibetan Plateau intensify this process, by introducing hot air much higher in the atmosphere.

The moisture-laden winds, which otherwise would have made land-fall in the Horn of Africa, find themselves yanked back on themselves (the Horn receives its rains either side of the summer monsoon season, when the trades go back to normal). As the winds rise up over India's great plateau, they are forced to drop their moisture; but as the rain condenses out, a runaway effect is created because condensation releases yet more heat, the 'latent heat', which pays back the extra energy needed to evaporate the water in the first place (which is why fountains feel cool: the evaporation they induce absorbs heat from the surroundings).

To create a monsoon, therefore, all you need is a strong heat contrast between land and sea, and a source of moisture-laden air. At times in the Croll–Milankovich cycle when seasonality is most pronounced (say, a Southern Hemisphere Summer coinciding with the Earth's closest approach to the Sun) any monsoon in that region will be enhanced. And that is the sort of cyclicity that geologists expect to find when looking at the rocks laid down through many such cycles over thousands of years.

Computer climate models for a Pangaean Earth show that by far the greatest portion of global rainfall in that time was convective, and took place at the Equator – and hence almost entirely over the ocean, Panthalassa and its reef-fringed embayment, Tethys. As the great 'C' of Pangaea drifted slowly northwards, coming to straddle the Equator more symmetrically, the huge baking landmasses now sitting just north and south of the Equator exerted a gigantic effect upon the distribution of rainfall.

Tethys, embraced by the supercontinent, was a warm ocean. An equatorial current flowed directly into its maw, concentrating heat and nutrients gathered from Panthalassa and introducing massive amounts of moisture into the atmosphere above it. However, the huge land areas north and south of the gulf would have set up Northern Hemisphere Summer monsoons on Tethys's northern coast, and a Southern Hemisphere Summer monsoon on its southern flank. Climate modellers believe that this effect dwarfed even the biggest modern monsoon, and have dubbed it 'megamonsoon'. Also, Pangaean mountain belts were probably among the mightiest ever seen on Earth. Those bordering Tethys's northern coasts would have mirrored the enhancing effect of the modern Himalayas on the modern Asian summer monsoon, making the megamonsoons even more so.

A new Lyell

In conjuring these vanished worlds back into being in such great detail, geologists use two forms of uniformitarian reasoning. They project physical constants back into the past (adjusting for secular change, such as the Sun's slowly increasing energy output) because physical laws do not change with time. And they interpret sediments in the light of what is known, by inspection, of sedimentary processes and environments around us today.

Working from several lines of evidence (including fossil magnetism in rocks, fossil animals and sediment types), geologists can determine where all the broken bits of Pangaea used to be and how they fitted together, giving a broad outline of the supercontinent. On to this outline, the ancient topography (young, high mountain belts like the Urals, older ones like the Pennines, the basins and plains) can be plotted. Those fossils and sediment types that give firm indications of climate – so-called 'climate proxies', such as glacial deposits or coals – can then be added to the picture. If the geologists have plotted and interpreted the rocks and fossils correctly, if the assumptions made about them by analogy with modern sediments and living things are correct, if the palaeomagnetists have got the continents in the right place, if the modellers have understood the palaeoclimate correctly, and if the computer model is truly reflecting the way energy balances between land and sea and the way oceans and the atmosphere exchange heat and moisture, then everything should fit perfectly and make sense. Needless to say, it rarely does, and this is what keeps it interesting.

To objectify the process of deciding if the distributions really do make sense – to make it more 'scientific' – modellers compare the geological evidence (often referred to as the 'ground truth') with computer predictions generated by (more or less) the same sort of computer models used every day to generate weather forecasts. These massive programs attempt to mimic the complexity of the Earth's climate system by breaking the hydrosphere and atmosphere down into layers and the geography of the Earth into manageable pixels 1.25 degrees square. With a supercomputer doing all the calculations, they re-create ancient water temperatures, winds, evaporation, cloud cover, storminess, snow depth, soil moisture, hurricanes, monsoons. The lost continents of science are brought to life partly in machines.

Energy Balance Models look at the land–sea distribution and solve the thermodynamic equations that can give some idea of how hot the land was relative to the sea at different times of the year. Climate-predicting programs are called GCMs, General Circulation Models, and combine fluid dynamics with ancient geography to simulate the climatic response to the energy balance. GCMs that try to simulate the atmosphere are called AGCMs, while OGCMs treat the circulations of the ocean. In recent years these have been brought together in coupled ocean-atmosphere circulation models (OAGCMs). Researchers can tweak the parameters of these models – for example, to take account of the Sun's lower energy output 250 million years ago, or to allow for different mixes of gases in the air at different times in the Earth's history. They keep tweaking until the model matches the evidence – or exposes anomalies that merit closer inspection.

Model predictions of the Pangaean megamonsoon have one major thing in common: all predict strong seasonality on and around the northern and southern shores of Tethys. And seasonality is something that geologists can look for, because pronounced seasons leave behind patterns in sediment sequences. Also, plotting particularly climate-sensitive rock types on a reconstructed map of the supercontinent will produce a pattern that – if the monsoon phenomenon is real – will not be perfectly zonal, as might result from the atmospheric cells alone. The monsoons will perturb this pattern, and the zones will depart from perfectly paralleling latitude.

In orbit over Pangaea

So, as the waters seep in, the Zechstein Sea fills and the drowned dunes release their sudden frothy exhalations, let us avoid the possibility of being surprised by a gorgonopsian, the top predator of its time, and soar through the air in which no bird has ever flown, up through the circulating atmosphere to the edge of space, and look down upon the latest (but not the last) supercontinent.

Below the bands of cloud, Pangaea sits within the globe like the 'C' in a copyright symbol. The curve of land encloses a great sea – an inland ocean, the Tethys – whose east-facing opening to the global ocean Panthalassa is partially obstructed by a number of small island subcontinents covered in dense jungle, much like Borneo or Sulawesi today. One day these microcontinents will drift north and collide with the northern limb of Pangaea to form much of what is now China. But for the time being they are the only major land areas not accreted to the supercontinent, which is just now at or about 'maximum packing'. Mountain building is still taking place along the northern shore of Tethys, which is fringed by a long mountain range created by the subduction of Tethyan ocean floor and the occasional accretion of those small continental fragments, waiting like ships outside harbour. This line of mountains already includes the older, northern ranges of the great mountain belt most people refer to collectively as the Himalayas: the Tien Shan and Nan Shan mountains.

The northern limb of Pangaea, stretching from Siberia, the Urals, Europe, Greenland and North America to the future Pacific rim, is known as Laurasia and was formed when the Ural Mountains were raised in the collision of North America with Europe and Siberia. This towering young belt bisects Laurasia north–south, to the northern Tethys shore. To the west the Hercynian (and beyond them the older Caledonian) mountains stand proud; but they are much older and (thanks to millions of years of erosion) already less pronounced. Between these ranges a finger of sea reaches south from the Boreal Ocean and feeds a growing area of water, the Zechstein Sea, that will soon spread south and east to cover much of future central Europe, bringing moisture to the heart of the great northern deserts. On the other side of the Pennines another inland sea, the Bakevellia Sea, fills a basin that mirrors the shape of the modern Irish Sea.

From its desolate western shore, the lone and level sands are interrupted only by the Appalachians, the US continuation of the Caledonian range, before vast stretches of sandy and rocky desert extend for thousands of kilometres towards what is now much of central and western America, where shallow ephemeral shelf seas and massive reef complexes mirror, on a much larger scale, what is happening in northern Europe.

To the south, where the supercontinent narrows to its equatorial waist, the Hercynian mountain system cuts inland, rising to four kilometres above sea level and marking the suture between the continental blocks of North America and North Africa. The range separates Laurasia from Pangaea's southern lobe, Gondwanaland. At its western end, where it meets the longitude-parallel Panthalassan coast, it turns south, defining the coastline of the future South America: the early Andes. At their southern tip these mountains touch also the Cape of South Africa, skirt Antarctica, and run up the western coast of Australia before coming to an end on the south-eastern extremity of Tethys, and so completing our round-Pangaea trip.

Gondwanaland is a much more ancient entity than Laurasia, and many traces of the older suturing events that brought it together can be seen in the remnants of much older mountain ranges, one of which runs between eastern South Africa, Antarctica and the eastern coast of India, passing through 'Gondwana Junction', where those three future separate continents now touch. There, in 250 million years' time, when the old sutures have opened up again, thousands of miles of ocean will have squeezed into the crack, and the Vivekananda Memorial will stand on a rocky islet of charnockite at India's Land's End, staring out across the sea to its vanished neighbours.

Forests of the polar night

As Pangaea has moved steadily north through the Permian, the South Pole has all but slipped into the sea. The great continental icecap that had existed for many millions of years since the Carboniferous has finally melted away completely, releasing the last of its cargo of mud and boulders the size of men. Unique to Gondwanaland, dense forests of *Glossopteris* trees, standing up to twenty-four metres tall, the shape of Christmas pines and growing a thousand to the acre, fringe the southern coasts of Tethys and stretch inland to within twenty degrees of the pole.

These forests of the polar night withstand two seasons: one of feeble light and one of unremitting dark. Today's world has no equivalent of this eerie ecosystem. Their growth rings show that each summer these trees grow frenetically. Those nearer the coast are lashed by megamonsoon rains roaring in from Tethys, the thick cloud further weakening the feeble sunshine raking the latitudes at the bottom of the world. And as the brief growing season comes to an end, and the orbital progress of the leaning Earth draws the sun in its undulating course daily closer to the horizon, the tongue-like leaves turn wild and fall on thick beds of countless others on the sodden forest floor. The sun dips further, finally no longer peeping above the ending line, and all growth ceases for six months without prospect of a dawn.

Leaves that will one day lie fossilized beside the frozen body of Captain Scott fall into the anoxic peat. The great Permian coals store up the Sun's ancient energy like a battery, waiting for release in power stations and steel mills.

These coal-producing forests occupy a climate zone designated 'tropical everwet' and, according to the occurrence of coals at this time, this zone extended from about midway along the southern coast of Tethys, across the island archipelagoes standing in the great gulf's mouth, to the northern shore's eastern promontory, and then back west, ending not far short of where the Ural mountains join the coastal cordillera. Oil source rocks and coral reefs cluster here, bearing testimony to the high organic productivity of the Permian tropics.

Around the reef-fringed Tethys, only rarely does this everwet zone give ground, and then mostly to 'tropical summerwet' conditions that also prevail across the mountainous mid-section of equatorial Pangaea and extend only a little way east along each shore of the great embayment.

Tropical summerwet is too dry for coals, and none is found today in places that were once situated here. But coal can, and does, form beyond the everwet tropics. It is a common misconception that all coal forms in steaming swamps like the Amazon or Congo basins of today. The main requirement for coal formation is a high water table that prevents plant matter from decaying. So, if that can be combined with high productivity of plant matter, coals can also form in cool and warm-temperate climate zones. This was particularly true over Gondwanaland, clothed with its unique *Glossopteris* forests, growing amid the lakes and valleys of the sodden, recently deglaciated southern continent. But plant remains (not abundant enough to make coals, though significant enough to create tantalizing 'floral localities') also extend around the shallow seas running south from the northern Boreal Ocean – like the Zechstein.

Despite these exceptions the main signature of Pangaea is one of almost unremitting aridity. The continent is too large for the moisture of the oceans to reach its interior; the late-Permian atmosphere, richer in carbon dioxide by perhaps five times the modern level, holds in the heat of the weaker Sun. In the parched heart of the northern and southern lobes of Pangaea summer temperatures soar over 45°C, while at polar latitudes they fall in winter below −30°C. South of the equatorial mountains, salt flats, gypsum playas and dune fields link the west coast of Pangaea, across the whole of the landmass that is now split into North America, North Africa and Arabia, to the southern shore of Tethys. North of it, desert; from shining, reef-fringed western shores bordering Panthalassa, all the way to the towering Urals – interrupted only by ephemeral, evaporating seas, recently filled and soon to teem with rich, spiny shellfish adapted to their bitter, hypersaline waters.

Further north, around the Boreal Ocean and its embayments, under the northerly storm track, the prevailing westerlies bring moisture in from Panthalassa, just as they do today from the Pacific to the boreal rainforests of moss-curtained pinestands of Washington State and British Columbia. But that moisture is soon spent and cannot penetrate far inland, so these conditions give way rapidly eastward to cool and finally cold temperate zones along the chilly, but ice-free, roof of the end-Permian world; a silent tundra shore, where soon some of the largest volcanic eruptions in Earth history will devastate hundreds of thousands of square kilometres, burying them in kilometres-thick lava piles and nearly bringing the whole story to an end, even before the first dinosaur stalked the planet.

World reborn

Since Alfred Wegener first pieced it back together in 1912, Pangaea continues to be reborn in the minds of Earth scientists – and their computers – as a living and (just about) breathing world; a unique place with many lessons to teach us about how our planet's climate works. It is the supercontinent about which we will always know most, because it is not long gone; its sediments are everywhere; our modern oceans contain a magnetic road map that helps us reconstruct it from its shattered remnants. Pangaea gave us much of our coal; Tethys laid down most of our oil and gas; evaporites formed in its shelf seas gave us nearly all our salt, on which almost all our chemical industries were established. The Zechstein Sea even gave us the fabric of the Palace of Westminster, and the treacherous landscape of Ripon. It even gave us dinosaurs, Alice and the rabbit hole.

Pangaea was the first lost supercontinent that actually *existed* to be imagined by the human mind. In one sense, it is still and always will be a fantasy; but one constrained by uniformitarianism – not Lyellian, but one that truly takes full account of the importance of the rare event in geological time. Thus the human imagination is held within the fruitful confines of method. You can see the effect of this in every academic reconstruction of Pangaea. Ever since Wegener himself, whose didactic purpose made them necessary, the outlines of the *present* continents are always made clear, embodying the claim of this supercontinent to its objective reality.

But what of older supercontinents? What of the supercontinent that broke up to give us Pangaea? And the one before that? Compared with Pangaea, those lost worlds seem truly lost. As with all geological evidence, the older it is, the less of it survives, the more mangled it has become and the harder it is to interpret. It is all but impossible to picture them – to see oneself standing on them – as you can with Pangaea. They have their magical names, which lend them reality of a sort despite the fact that, for some, even their very existence remains controversial. About Rodinia, Pannotia, Columbia, Atlantica, Nena, Arctica or distant Ur, the mists of time gather ever more thickly.

7

WORLD WARS

At a specified time the earth can have just one configuration.
But the earth supplies no information about this. We are like
a judge confronted by a defendant who refuses to answer, and
we must determine the truth from circumstantial evidence . . .

ALFRED WEGENER

Freikörperkultur

A white-backed vulture, circling high over the empty central deserts
of Namibia one day in 1940 would have seen something odd going on
at the bottom of one of the rocky gorges of the Kuiseb River, which
drains the Khomas Hochland west of Windhoek. Unusually for a
Namibian river, the Kuiseb does not peter out into the Namib Desert
but runs into the South Atlantic at Walvis Bay, the major port along
Namibia's beautiful but forbidding Atlantic coastline.

There has been a river valley at Kuiseb for perhaps thirty million
years, though the present gorge is as little as two million years old,
dating from the beginning of the last Ice Age, when global sea levels
fell dramatically. Today, for most of the time, there is no water in the
Kuiseb River, but the size of some boulders, the smoothing of the
rocks high along its banks, or the occasional telltale tangle of logs and
brushwood, lodged way up in the cliff, speak of the river's terrible

force when rains finally come to Khomas Hochland. Yet, luckily for the game, and the occasional bushman trekking by, water often persists in isolated pools on the valley floor, even through the dry intervening months and years.

At any time such water holes might play host to a troupe of zebra, encircled by a cloud of hoof-kicked dust; or to a lone gemsbok, his dark, ribbed horns sweeping upright as he dips his head to drink in the still heat. Light-coloured tracks lead off from the pool in all directions. All around the water the trampled dust lies like buff flour. There is no patch that does not bear a hoof print. An animal reek, from the spoor and urine with which the visiting beasts pollute their source, hangs in the air. But on that day in the first year of the Second World War, the hole has human visitors.

Two naked young German geologists are wading through its tepid, greenish water, catching carp with a makeshift fence-net made out of two bed sheets sewn together with a pair of string underpants and stiffened with tamarisk twigs. One of these intrepid hunters is Dr Hermann Korn; the other – owner of the sacrificed pants – Dr Henno Martin. Both are on the run.

The two fugitives, who had both studied in the ancient University of Göttingen in Germany, had rejected the rise of fascism in their own country and emigrated to the protectorate of South West Africa, a former German colony. They earned their living on water exploration projects and together got to know the geology of this remote country. But not remote enough; before the growing tide of war, the two men, fearing internment by the South African Mandatory Government, hatched a plan. They would escape, to live a Robinson Crusoe existence in the desert they had come to love; and on 25 May 1940, with a stolen truck full of essential provisions, their dachshund Otto, an air rifle, a pistol and some ammunition, they set off for the wilderness.

Their desert sojourn, a constant battle for survival fought over water, food and the many dangers of the desert and isolation, lasted two years. It was described in a book known in English as *The Sheltering Desert*, which Martin wrote for his wife and published in 1956, ten years after Hermann Korn was killed in a road accident.

Martin and Korn were not on the run for their lives. In the desert their lives were in just as much danger as they would have been in an internment camp. Yet – and Martin's book does not make this clear, perhaps because it would have been too painful a thing to say in post-war Germany – the prospect of internment was not made so appalling by the fact of imprisonment alone, or even fear of the regime in whatever camp might have received them. It was fear of their fellow internees, who would most likely have been ardent Nazis.

Forsaking the hideous regime in their native Germany for the freedom of the southern African desert in the late 1930s, Martin and Korn found themselves in one of the bastions of free tectonic thought and quickly became aligned with it. Here, where geological evidence for the break-up of Gondwanaland was at its most stark and undeniable, Wegener already had his greatest champion. In the heart of Suess's old imagined domain, amid the reality of evidence that the great London-born Austrian had first pieced together from books in Vienna, Henno Martin and Hermann Korn found themselves walking in the footsteps of one of the greatest field geologists of all time: Alexander du Toit.

In 1961, just as the tide was beginning to turn for Wegener's theory, Henno Martin delivered a memorial lecture to Alex du Toit at the Geological Society of South Africa, appropriately entitled 'The hypothesis of continental drift in the light of recent advances of geological knowledge in Brazil and south west Africa'. The paper was a development of Namibia-based work he and Korn had carried out (partly during their desert exile) and published in the decades before;

gathering detailed evidence of the ancient glaciation in that geologically unknown region, and fitting it together with the patterns of
glacier movement first assembled by Suess and later taken up by
Wegener in his reconstruction of Pangaea.

Martin's later work in Brazil, which sought to make the correlations between sequences now separated by the South Atlantic more
precise and therefore persuasive, followed pioneering work in the
same vein by du Toit. South Africa's greatest geologist, a man decent
and truthful to the roots of his hair, came in 1924 to be on another
continent pretending to the Carnegie Institution of Washington,
which was paying his expenses, that he had nothing more on his mind
than the simple collection of data whereas in fact he was looking for
evidence to prove a rogue theory that most geologists (and especially
geophysicists) in the USA ridiculed.

The 'colonist'

Alexander Logie du Toit was born in 1878 at Klein Schuur, under the
shadow of Devil's Peak and Table Mountain, in the Colony of the
Cape of Good Hope. His protestant family, which originated around
Lille in northern France, began its journey to South Africa from the
Netherlands, where it had been driven by religious persecution. From
there in 1687, two brothers du Toit sailed for the Cape, establishing
their family dynasty one year ahead of the main body of Huguenot
settlers, who contributed so much (include wine-making expertise) to
South Africa. Du Toit eventually became one of the most widespread
family names among the province's settlers.

The other element in du Toit's genetics was Scottish. Shortly after
the British took over the Cape, Alexander Logie, a naval captain from
Fochabers, Inverness, married into the family and took over the estate
at Klein Schuur. Scots genes, however, had a little more difficulty

200 Ma Novopangea

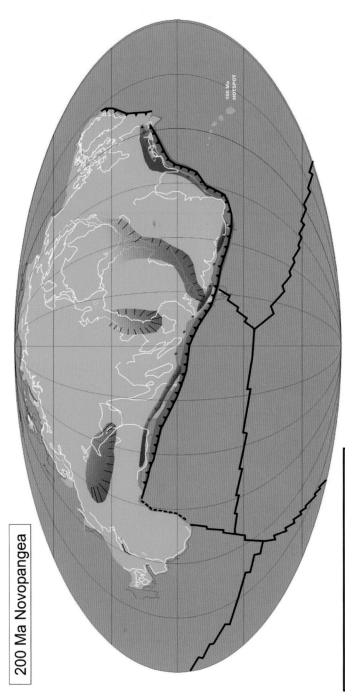

100 Ma
HOTSPOT

KEY

∘ Volcano

ᵛ Volcanic margin

Land areas

Shallow seas

Plate boundary
(constructive)

Plate boundary
(destructive)

Plate boundary
(conservative)

Mountain ranges

Preconstruction by Roy Livermore for John Adams Television

22-2-2006

'Novopangaea', the future world envisaged for 250 million years hence by Dr Roy Livermore for the BBC television series *The Future is Wild* (see Further Reading). Novopangaea is the product of 'extroversion'. Reproduced with permission, Dr Roy Livermore.

Dr Henno Martin (1910–98) at 'Carp Cliff', beneath which he and Hermann Korn had spent much of their time in hiding. Photo courtesy of Dr Gabi Schneider and The Geological Survey of Namibia.

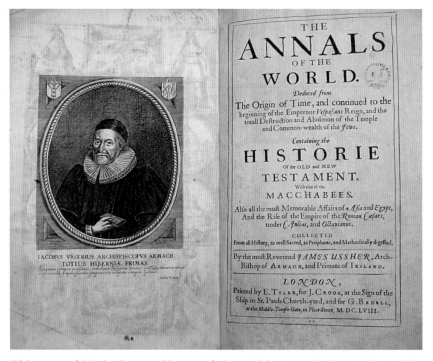

Title page of Ussher's great history of the world. James Ussher (1581–1656), Archbishop of Armagh and Primate of Ireland. Ussher set himself the task of analysing astronomical cycles, historical accounts, and several sources of biblical chronology, to determine the precise date on which God created the Earth. Photo by Ted Nield, courtesy of The Society of Antiquaries of London.

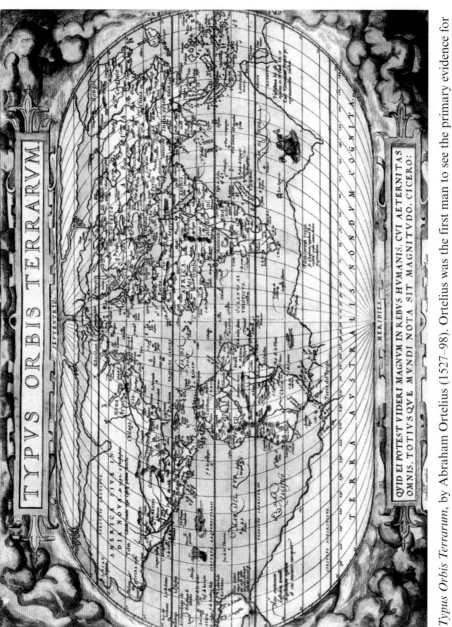

Typus Orbis Terrarum, by Abraham Ortelius (1527–98). Ortelius was the first man to see the primary evidence for the opening of the Atlantic, and the first to make that wild surmise.

Helena Petrovna Hahn
(1831–91), better known as
Madame Blavatsky.

Minnesota congressman,
Ignatius Donnelly (1831–1901),
author of *Atlantis: the
Antediluvian World* (1882).

Eduard Suess (1831–1914) as a young man. Reproduced with permission of The Geological Society of London.

Alfred Lothar Wegener (1880–1930). Reproduced with permission of the The Alfred Wegener Institute for Polar and Marine Research.

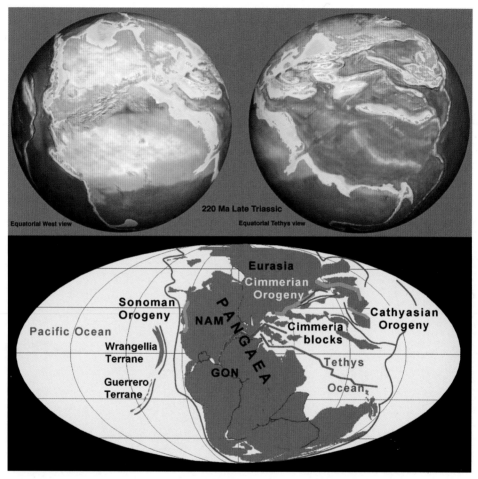

Reconstruction of the supercontinent of Pangaea, 250 million years ago.

Alexander Logie du Toit (1878–1948).

Dr Willem Anton Joseph Maria van Waterschoot van der Gracht (1873–1943). From *AAPG Explorer: A Century; The debate begins, Wegener drifts into gear, 1999*. Reproduced with permission of AAPG.

Professor John Tuzo Wilson (1908–93). Reproduced with permission of the Ontario Science Centre.

Professor Janet Watson (1923–85). Reproduced with permission of The Geological Society of London.

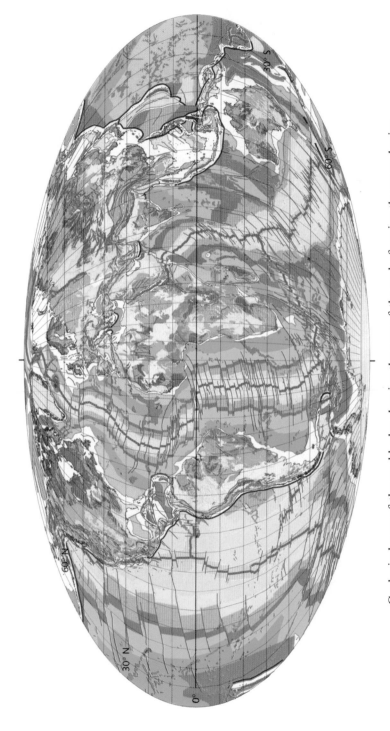

Geological map of the world, also showing the ages of the rocks forming the ocean basins.

gaining entry. Alexander, having no children of his own, adopted his wife's nephew (also called Alexander Logie du Toit) as his son and heir. He eventually came to marry *his* cousin, Anna Logie. Their union produced four children, the eldest destined to become South Africa's first and greatest home-grown geologist.

Du Toit graduated from the forerunner of the University of Cape Town and then attended the Royal Technical College in Glasgow, where, in 1899, he qualified in Mining Engineering. After two years studying at London's Royal College of Science, he was invited back to Glasgow, where he held two posts, as lecturer in Mining Engineering at the Royal Technical College and in Geology at Glasgow University. Then, in 1903, du Toit sailed back to the Cape to join its new Geological Commission, charged since 1895 with geological surveys and mapping in the resource-rich colony.

In the years between beginning work in 1903 and leaving the Survey in 1920, he mapped in detail over 50,000 square miles between the Cape and Natal. Of this enormous area, nearly 43,000 square miles found its way into published geological maps. Much of his work was carried out from a donkey wagon, his mobile home kept by his diligent and long-suffering Scottish wife. From this vehicle he would venture out across trackless mountain and veldt *on a bicycle*. And that is not all. Geological mapping depends on having a base topographic map – or today an aerial or satellite picture – on which the geologist plots his rock types and readings of technical measurements. Nothing useful can be done without one. However, for much of the area du Toit mapped, there were no such maps. So he made them too, using a small plane table (a level sighting and drawing surface mounted on a tripod) that he would carry with him on his bike. In effect he mapped much of this vast area twice.

There is always a tendency to beatify the departed in their obituaries, but reports of du Toit's character are too consistent for this to be the case with him. For all his stature and eventual fame among fellow geologists, du Toit was innately modest, eternally wary of lime-light, kind and generous with people he met, in whom he clearly took a warm and sympathetic interest whatever their status. Reports abound of his phenomenal powers of recall, he not only wrote every-thing down but remembered where he wrote it down. Yet, as with Suess, his attention to detail was matched by an equal flair for the grand vision. He was also an unstinting worker, whose superhuman capacity for physical and mental effort continued almost unabated right up until his death, from a cancer diagnosed two years before but which he bore in silence and in secret.

One would think that he must have been an impossibly saintly man, were it not for the sly twinkle in so many of his portraits and an ingrained liking for devilment, for upsetting the applecart. The first inkling of this trait would come from his air of faintly amused detach-ment, his quaint and subtle sense of humour and certain unexpected pleasures. He had an unnerving tendency in company to spring up suddenly and recite questionable limericks. His musicality extended to a lifelong love of playing the oboe; and at one stage this outwardly quiet and unassuming intellectual took up motorcycle racing. He cut a strange figure for a rebel, but rebel he was.

Back in the USA

If anyone needed an example of how a theory's acceptance could have benefited from a bit of simple PR, then Alfred Wegener's book *On the Origin of Continents and Oceans* provides it. The truth is that Wegener could not have done more to antagonize scientists in the USA if he had tried.

Derek Ager was able to write in 1961 that while he himself was (then) in a minority among British geologists in opposing drift,

'American geologists appear to regard the Declaration of Independence as retroactive to the Palaeozoic.' The hypothesis of continental drift was nowhere more despised and rejected than in the United States, and by the late 1950s it set the 'island continent' apart from the rest of the scientific world. This division today seems all the more remarkable because now you might be forgiven for thinking it was American geophysicists who invented it. The truth of how this schism came about is an object lesson in the influence of prevailing culture upon science; and how it was eventually healed, as the great supercontinent of science reunited around a new paradigm, demonstrates the self-correcting nature of the scientific endeavour.

Objectivity in science is a contentious and difficult issue. At its heart lies a basic question: how should a scientist approach nature? To take two extremes: one, you go to nature and record what you observe. You clear your mind of any explanatory theories and amass data. But factual data on their own do not explain anything, so after all this observational effort you allow explanations to *emerge*. This is called the inductive method. Alternatively, you can first develop a theory – a hypothesis – about how you think nature works, and then go out to test that theory by observation. This is called the deductive method, or sometimes the hypothetico-deductive method.

At their extremes these two approaches produce, in one case, colourless fact-gathering, and in the other, an unhealthy dominance of ruling theory that blinds the observer to observations that don't fit. Clearly, the way in which science *really* works is a sensible and pragmatic mixture of the two; but balancing them has always been fraught with difficulty, not least in geology.

Many early geological thinkers in the eighteenth century had deductive tendencies. They imagined all-embracing 'theories of the Earth', expounded in thick tomes that offered ideological frameworks of how the Earth evolved into its present state. By the time the

Geological Society of London was founded in 1807, after the spectacular failure of several such ruling theories, advanced thought had swung back towards the inductive method. The general mood of the time was that science needed more facts. The stirrings of an 'Anglo-Saxon' approach to science were beginning to make themselves felt. The great early theorists of the Earth had been chiefly French and German. And while the Scots geologist James Hutton (1726–97), originator of uniformitarianism, also called his great book *Theory of the Earth*, his approach was truly more aligned with the observers than the grand theorizers.

Not that Hutton lacked grand theory. It was his concept of judging the record of the past by extrapolating from the present that Charles Lyell developed into its extreme form in the nineteenth century; and thus gave the whole tenor of the research conducted and published by the Geological Society of London its objective, inductive cast. This was very much in opposition to the caricature of continental science as over-theoretical and dogmatic, driven by God-like professors with their ruling theories and personal authority. This caricature, like all good caricatures, has some basis in truth, and is reflected in the German term *Weltall* that is given to this kind of theory and which means 'whole world'.

Geologists in the USA, however, were facing other problems. First, there was the immensity of the continent – still largely unexplored – to be documented. Second, they remained deeply suspicious of European 'authority'. They knew that evidence about the way the Earth worked that derived from *their* continent should carry no less weight than that from any other part of the globe. Moreover, they preserved a revolutionary dislike of pronouncements from on high. Somehow their science also had to declare independence. They had to find a new way of doing science, a democratic method that eschewed discredited Old World ways.

What they hit on, first of all, was induction. But realizing that this cure could end up being worse than the disease, and recognizing that the collection of data alone was not what science was about, America strove to develop a third way of its own, a method called multiple working hypotheses.

Under this method the scientist, like a good parent dividing his attention equally and impartially among his offspring, first presented his factual findings in as 'theory-free' a way as possible, and then discussed the observations in the light of as many different explanations as seemed reasonable. If in due course a leading theory or two emerged, these must be further refined by the subsequent collection of more data. From the end of the nineteenth century this approach quickly became the accepted, democratic, American way of science.

Into the middle of this new, republican method of doing science, expressly devised to counterbalance that outmoded, system-based theorizing of the Old World, dropped Wegener's *On the Origin of Continents and Oceans*.

Blast from the past

Even today, with our modern obsession with 'directed' research, and when a scientist applying for research money won't stand a chance of getting any unless he or she appears to be testing some grand hypothesis, Wegener's book makes strange reading. But to Americans in the mid-1920s, which was when J. G. A. Skerl's translation came out there, it read more like an affront to decent society.

We have already met Skerl, the man who should strictly be credited with introducing the English-speaking world to the word 'Pangaea'. It is fair to say that his translation must share some of the blame for the revulsion that greeted Wegener's book in the USA. From its opening paragraphs, everything that was upright, noble and distinctive about American-style scientific method was, it seemed, being given a deliberate and blatant slap in the face.

The ideology behind American science was that it had no ideology. It built on the British model of facts first, interpretation later (if at all); but then went further. Science had to be democratic through and through. The multiple working hypothesis method enshrined these egalitarian motives in the way scientists did their everyday work. America prided itself on its pragmatic, no-nonsense approach, and on hard work.

Wegener's book, by contrast, seemed to hark back to 'the bad old days', as they would have seemed, especially to American readers.

'The first notion of the displacement of continents came to me in 1910 when, on studying a map of the world, I was impressed by the congruency of both sides of the Atlantic coasts,' Wegener writes in his first chapter. He goes on: 'This induced me to undertake a hasty analysis of the results of research . . . whereby such important con-firmations were yielded that I was convinced of the fundamental correctness of my idea.' In the English version of the fourth edition the words 'hasty analysis' are replaced by the less pejorative 'cursory examination' but the effect was the same.

Wegener in full flight often reads a little like the letter of a madman: an obsessive with no education in the field, who suddenly undergoes a Damascene conversion that sets him off on a selective spree looking for confirmations of his Big Idea and rejecting anything else. After this strident and frank opening, which could hardly have been better calculated to annoy Americans, there is the small matter of the over-use of the sensitive word 'proof'. Skerl used 'prove' and 'proof' indiscriminately to translate gentler German originals, which might have been better rendered as 'demonstrate' or 'evidence'. Wegener's apparent dogmatism, enhanced by Skerl, gave the impres-sion of an inexperienced, auto-intoxicated 'armchair' scientist with an *idée fixe.* Other 'working hypotheses' were not given a fair and equal chance.

Wegener's book, in other words, was a polemic. He knew he was right. And what is more, when the criticisms began to rain down upon him, instead of meeting them halfway in an attempt to win his opponents over, Wegener merely became more and more adamant. In return, the trenchancy of his prose excused (and even encouraged) the brusquest of dismissals.

As the eminent scholar Mott Greene has pointed out, much is made today of the fact that Wegener was not a 'proper' geophysicist and was therefore shunned as an interloper. This 'fact' arises, I believe, because quite a lot of the mud that was slung at him by his near contemporaries still sticks. The same goes, as we shall see, for the assertion by the same opponents that there was 'no plausible mechanism' for drift. Wegener's opponents were not just vehement: many were implacable.

Wegener was, in fact, no less a geophysicist than any of his contemporaries, he just happened to have written a book about the physics of the atmosphere and to hold a post in the only part of geophysics, meteorology, that had any money in it. Eventually he did move on to hold a chair in Graz, Austria, whose title had the word 'geophysics' in it. His writings, published in proper geophysical journals, show that he was fully conversant with all the relevant material, and his ideas were discussed, albeit rejected, by geophysicists everywhere. They could not ignore him. As Greene has written, if all this 'does not identify Alfred Wegener as a geophysicist then nothing can and we can all retire to bedlam'.

Cultural differences also play their part here. If the meaning of the word 'science' can be so completely different between the Anglo-Saxon world and everywhere else (where it means 'organized knowledge' about anything, including literature, for example), there is no hope of attaining worldwide agreement on the meaning of 'geo-physics'. In the 1920s the understanding in most languages of the term 'geophysics' as the physics of the 'solid' Earth had not yet come about, and it still hasn't in Russian. Indeed, in that language the term 'geophysics' still is understood to embrace all Earth sciences.

Cultural incongruities dogged Wegener. His PR skills were not good. His method of doing science was out of favour. The Great War had ended almost a decade before, but the mid-1920s were still not a good time to be German. Wegener himself also came with brand-name difficulties. As the eminent historian of science Naomi Oreskes has pointed out in her brilliant and comprehensive textbook *The Rejection of Continental Drift*, to his geological readers the name Wegener sounded 'eerily similar' to the name of his eighteenth-century compatriot Werner.

Abraham Gottlob Werner (1750–1817) was perhaps the archetypal grand theorist of an old school, and at that time one of the most blackened historical figures in the Anglo-Saxon panoply of scientific bad guys. He was a vivid example of European system-building folly: the proponent of a discredited *Weltall* theory who, being already slain, was only too readily re-slain by subsequent generations. He was, in fact, still as unfashionable as only the most recently fashionable can be. More subliminally yet, the name 'Wegener' even seemed to echo another megalomaniac, world-building German Romantic, Richard Wagner.

Perhaps the final straw was the fact that Wegener's book had been written by a man who had gained the leisure to flesh out its ideas while recovering from wounds sustained during his country's recent assault upon the free, democratic world. In 1914 Wegener was drafted, and was shot in the arm during the assault on Belgium. Two weeks after his return to active service, he was shot again, this time in the neck, and his days of active duty were over. The fact that this objectionable theory had come to maturity in the mind of a man invalided from the hordes of the Boche was tactlessly given away in the author's Introduction to the third and fourth editions. Drift seemed to pose as much a Teutonic threat as a tectonic one, with Wegener being described in print with such choice words as 'forgetful', 'selective', 'unscientific' and even 'deranged'.

In Britain things were less heated. At the British Association meeting in Hull in 1923, British geologists had met to consider the drift idea. According to a report of the 'lively but inconclusive' meeting printed in *Nature*, the main common point of opposition occurred over Wegener's idea that Pangaea could have broken up and the Atlantic opened during the Quaternary, the most recent geological period. Nobody was very sure at that time when, in years, the Quaternary had begun (actually about 1.6 million years ago) but, while they might have been prepared to concede that continents moved, they didn't think it likely that they moved *that* much.

But while British geologists were arguing over the speed of the process, the basic concept did not seem to give those present much trouble. It helped that many of them had seen the African and South American rocks, but there was neither philosophical outrage nor any feeling at all that public decency had been affronted.

But there was another problem facing Wegener in the USA that did not affect his Hull audience. British geologists thought of isostasy in terms of the Airy model, with its mountain roots, that presupposed that rocks not only could flow but did. This was not the prevailing view in America. There a different model held sway.

Models versus reality

After a seventy-three-second flight, at 11.39am on 28 January 1986, Mission 51L, the Shuttle programme's oldest vehicle, *Challenger*, and its seven-strong crew, were blown to oblivion in the Florida sky. The brilliant physicist Richard Feynman quickly and intuitively hit on the immediate cause. Standing on the launch pad in freezing weather had reduced the springiness of rubber O-ring seals between the sections of *Challenger*'s solid rocket boosters, causing them to fail under pressure during lift-off. In a press conference Feynman demonstrated this simple idea using a section of O-ring, a beaker of iced water and a simple clamp.

But that engineering failure was only the immediate cause. Subsequent investigations exposed many quality-control shortcomings at NASA, notably a tendency to tweak the safety envelope rather than the engineering, so that systems could be certified 'safe' more quickly and so allow NASA to meet its tight launch schedule, dictated largely by the need to put secret defence-related satellites in orbit. In his appendix to the Rogers Commission report, Feynman reserved his most withering criticism for this. NASA, he said, had made the fundamental scientific mistake of confusing their models with reality. 'Nature,' Feynman wrote, 'cannot be fooled.'

Models encapsulate our view of the world according to scientific principles and hypotheses, though, like supercontinents, even those that really did once exist, they are no more than imaginary constructs. Models are maps of reality, and like all maps, they have errors and it's never too long before someone finds something that was missed.

Nevertheless, it is all too easy for scientists to forget this and start thinking that model and reality are the same. As we have seen, in Britain the accepted model of isostasy – the theory that explains why mountains stand high and oceans stand low – was based primarily on the Airy model. This envisaged mountains as the tops of great rocky icebergs with deep 'roots', 'floating' in the Earth's mantle. But the other explanation, put forward by Archdeacon Pratt, supposed instead that mountains stood high because they were less dense and had no 'roots' reaching deep into the Earth. This idea, though not entirely mistaken, is less in accord with the true internal geometry of the planet than Airy's. However, Pratt's model does have one big advantage to mapmakers: it makes the sums easier.

If you belonged to a hard-pressed survey department in the early twentieth century, desperately churning out accurately triangulated maps of an emerging nation with a booming economy, you also had to correct thousands upon thousands of survey measurements for the effects of local gravity variations. And because nobody had yet invented the computer, this was all done manually, by armies of careful ladies with pencils. The big advantage of the Pratt model is that it makes these laborious calculations much easier. Easier means quicker, and time is money.

But it appears that the sober and practical US Coast and Geodetic Survey had become so attached to their convenient calculating strategy that some of them actually came to believe it was true, and none more so than the chief geodesist, William Bowie (1872–1940).

It is often said that the Airy model, with its implicit presumption that rock must flow over long periods of time, actually demands that continents can move. The Pratt model, on the other hand, carries the reverse implication. If you choose to believe it, then the idea of laterally sliding continents becomes worse than unnecessary: it becomes inconceivable.

In Pratt's model the base of the 'crust' is envisaged as a smooth, uniform surface sitting at the same depth everywhere. Mountains stick up above the general surface level to the extent that their rocks are less dense; the higher the mountains, the less dense the underlying rocks. But lateral motions of the crust would even out these density differences in time, and although the Earth is very old, we still have mountains. To a Pratt believer, therefore, drift seemed to be impossible. The influential Bowie had glimpsed an elegant model and taken her for Mother Nature: just because the maths was simpler.

Wegener was walking into a tough room.

Lone drifter

Not all US geologists were opposed to drift. Today (and indeed also in older literature, when American pro-drifters were trying to make it seem less un-American) continental drift is often referred to as the Taylor-Wegener hypothesis to recognize US Geological Survey geologist Frank Bursley Taylor (1860–1938), who proposed a form of drift in 1910. There were also others. But despite converts (many highly respected), it was slow going. As the Wegener theory's huge explanatory power won new adherents elsewhere, by and large, and for a wide variety of reasons (some spoken, others not), American geologists stood out against it almost to a man. Just how unanimous this opposition was became evident at a landmark scientific meeting in 1926 organized by Dr Willem Anton Joseph Maria van Waterschoot van der Gracht (1873–1943).

Three years earlier Royal Dutch Shell had fired this brilliant, Amsterdam-born geologist. Van der Gracht was by no means the last geologist to find that being sacked by a multinational was the making of him, and a distinguished career in the United States as a pioneer in prospecting by seismic interpretation lay before him. He was also destined to play a pivotal role in the history of continental drift theory. In 1917 he had co-founded the American Association of Petroleum Geologists (AAPG), today the world's largest Earth-science society, and in 1926 convened one of that body's meetings. The conference, in New York City, had a single purpose: to discuss theories of continental drift.

For the AAPG, which was then still a fledgling organization, this was a decisive moment. Its co-founder had chosen a controversial topic because it was, he believed, the biggest scientific game in town, with massive implications for oil prospecting. What better launch for a prestigious series of published symposium proceedings than to treat such a revolutionary subject?

In the mythology of the subject this meeting has often been portrayed in something of the manner of the famous confrontation at the British Association in Oxford in 1860 between Darwin's defender and champion, Thomas Henry Huxley, and Bishop Samuel Wilberforce. Although the earlier debate is always held up as a triumphant rout of the forces of fundamentalism, the New York meeting on drift is portrayed as one man's lonely defence of Wegener's theory against seemingly insuperable opposition. Neither of these histories is really accurate.

Wegener was not present in New York, and it should not be forgotten that his theory was by no means the only drift theory in town. Other theories about how continents might have drifted were also on the agenda; it's just that Wegener's happened to be the main target. Frank Bursley Taylor, for example, also presented his theory. The American survey geologist suggested that two supercontinents, formerly at or near the Earth's poles, gradually drifted towards the Equator, pushing up mountains along their leading edges and splitting under tension along their poleward, trailing edges. However, although Taylor's theory had been around for about two years longer than Wegener's, only Wegener had written so persuasively and at such length about it, and almost none of the other speakers addressed themselves to Taylor.

In fact, with Taylor concentrating on his own ideas, the only person present who was at all in favour of Wegener was the convener himself. Almost without exception, Wegener was condemned, and even those who did not join in the condemnation merely urged further research. Reservation of judgement, however, was not to be the order of the day.

The interesting thing about discussion meetings is, of course, discussion. If unanimity should unexpectedly break out, people reading the resulting proceedings will wonder why the meeting took place at all. So, in order to save his publication, and introduce at least a semblance of balance, van der Gracht solicited a number of supporting contributions from people who had not been present in New York. He commissioned contributions from Alfred Wegener himself, from the Irishman John Joly of Trinity College, Dublin, his fellow Dutchman Gustaaf Molengraaf, of Delft's Institute of Technology, and Glasgow University's professor John W. Gregory.

These men were known to be broadly open to drift theory. Joly (1857–1933) was a highly original thinker with his own theory. His contribution, though, barely covered two sides, and concentrated on generalities. Molengraaf pronounced himself a drift enthusiast, but quibbled about Wegener's insistence that Pangaea's fragmentation had all been westerly. Gregory was also lukewarm. He was not opposed to drift *per se*, but saw no reason why the present distribution of land and sea could not equally well be attributed, *à la* Suess, to vertical rather than horizontal movements.

These endorsements, though hardly ringing, did help, but not very much. So the convener decided to exert his privilege not only to have the first word, reproducing a seventy-five-page set of opening remarks, but also the last, book-ending the proceedings with a twenty-nine-page summing-up, in which he critically examined (and often demolished) various objections. In the end van der Gracht himself wrote 43 per cent of the entire volume. The proceedings were saved (and are now a sought-after collector's item); but in its list of contributors the transatlantic divide over continental drift theory was made flesh. The supercontinent of science had undergone a geological rift.

Du Toit's proof

In the pragmatic, fieldwork-obsessed world of early-twentieth-century American geology, where hard work and first-hand familiarity with the rocks were the only justification for any excursion into theoretical territory, the accusation of being an 'armchair geologist' was a grave one. It was levelled at, for example, Suess, whose great insights were indeed the product of much book learning and relatively little fieldwork (at least in his mature years). Both Suess and Wegener, in pointing to the geological correlations between the modern fragments of Gondwanaland, were generalizing from observations made by others about the congruence of facing Atlantic coastlines, Wegener doing little more than taking Suess's data and interpreting them differently.

American scientists might not have liked the method, but, some asked, what was the truth of this assertion? How alike *were* these sequences of rocks? Would it not be scientific to go there and check it out? The logic seemed faultless, but there was a snag. Proceeding in this way would be engaging in *deductive* research – setting out explicitly to test a ruling theory – and that was not the American way. Given that, who in the States would be prepared to put up the funds needed to send someone halfway across the globe to test a theory that most geologists in America had already dismissed?

By good fortune, Reginald Daly, Harvard University's Sturgis Hooper Professor of Geology, had been on a nine-month fact-finding mission to Africa in 1922, the year Wegener's book was published for the first time in America. The main focus of the expedition had been South Africa's massive Bushveld Igneous Province, and during the trip Daly had met Alex du Toit. The party had also included Molengraaf and Frederick Wright (1877–1953), a geologist from the Carnegie Institution of Washington, which funded the trip.

Returning that autumn a convinced 'drifter', Daly began to hatch a plan with his friend in that well-endowed institution. The result was an illuminating exchange of letters between Wright and the Carnegie's Director, John Merriam. In the first, Wright told Merriam about how the geological world was now in the thrall of an amazing new theory. He told his boss about the British Association for the Advancement of Science, and how its meeting in Hull had listed some crucial tests that should be done to verify the claims made for the geology of facing shores. The correlation of the Karoo rocks of South Africa with their South American equivalents particularly cried out for attention. There was only one man for the job, Alex du Toit. The Institution should pay for him to go to Brazil and verify it. In high hopes he sent off the proposal.

Merriam turned it down. After the initial shock of disappointment, however, Wright found that the boss had not closed the door completely. Merriam was not about to put to his Trustees a request to spend the Carnegie Institution's money on any overtly deductive search after facts. No, the proposal had to be more circumspect, more pluralistic, more American. To judge by Wright's second (successful) attempt, it also had to be less exciting.

Where version one had begun, like a piece of contemporary journalism, 'The hypothesis of continental drift . . . is arousing the keen interest of geologists the world over', version two opened: 'In accordance with your request for information I take pleasure in presenting, herewith, an outline of some of the results that may be expected from a comparative study by A L du Toit of the Karoo, Gondwana and other equivalent formations of the southern hemisphere . . .'

When Merriam finally gave the project the green light, Wright copied both versions to du Toit and explained the situation. Sure, *they* all knew why he was going to Brazil: to test Wegener's hypothesis. They just couldn't say so. This was important information for du Toit because, whatever his own intentions, he would now have to write up his results the American way, using multiple working hypotheses. This thoroughly deductive expedition had to be turned on its head and made to look like induction in action. The stated main purpose of the trip would be to gather facts: to add these good, solid bricks to the great edifice of science, and then to offer many different explanations of them, drift being merely one theory among equals. Wright's letter arrived in Pretoria the very day du Toit set off on the Trans-Karoo railway for Cape Town, where he would catch a boat for Brazil.

It would take du Toit two years to complete his monograph, and it would be four years before it finally appeared in print, in 1927. It is a curious document, and the tension between the dictates of American-style induction and du Toit's own pro-drift convictions shows in almost every line. Du Toit succeeded in confirming the transatlantic correlations of Suess, expanded by Wegener, and he added some more of his own; yet to his dismay, all his work hardly seemed to help the cause of drift. In fact, he often found his painstaking fieldwork dismissed as irrelevant or inconclusive; he was even tarred with the same brush as Suess and branded an armchair geologist for his pains – he, a man who had mapped 50,000 square miles!

Non posse

The reason for du Toit's failure lay in the nature of the evidence itself, of the 'proofs', as Skerl would have translated it. Both du Toit and Wegener were aware of this problem, which was that by their nature, geological correlations did not *compel* drift. In fact, du Toit committed a PR blunder of his own by admitting in his monograph that the similarities between the fossils of Africa and South America 'can generally be explained equally well, even if less neatly, by the orthodox view that assumes the existence of extended land bridges . . .'. This was putting weapons in the hands of his enemies; but his point was that fossils *on their own* could not decide anything because fossils are remains of living things and living things can move. Instead, he concentrated on the rocks themselves.

Sedimentary rocks change their character from place to place, depending on the environment that lays them down. Beach sediments change as the bay merges with, say, a river estuary. What du Toit showed was that sediments of the same age in South America and Africa often showed greater changes *within their outcrops in those continents* than they did across the wide expanse of the South Atlantic. This, for him, was far more compelling evidence than the fossil similarities that the two continents were once close together and hence probably joined along their now distant coastlines.

But this evidence was also circumstantial. Geophysicists persisted with their *non posse*, 'it's impossible', line of argument (it is a well-known fact that a scientist always finds evidence from his own field most convincing). If you wanted to win over physicists, it was little use drawing their attention to fossils and sediments. For that group, drift didn't happen because it *couldn't* happen.

Writers of popular science are often accused of 'hardening up' stories to make them simpler, clearer or more exciting. This sometimes gets them into trouble with scientists, who tend to favour caution. Charles Ray was writing *The World of Wonder,* my father's unwieldy science encyclopaedia, in the early 1930s. Wegener, referred to throughout as 'the late Professor Wegener', had not long perished on the Greenland icecap. Continental drift theory had been around for eighteen years but was still highly controversial. Yet undaunted, after an excellent summary of the theory, Ray writes: 'It seems a startling theory to think of the continents sliding or drifting over their foundations; but distinguished geologists say there is *nothing at all impossible in the theory from a mechanical point of view . . .*'

When I first read this as a boy, of course I believed it. When I came across it again later, having by then read the conventional textbook histories of plate tectonics, it made me laugh. Surely, the situation was quite the reverse? Was it not precisely the 'mechanical point of view' (the mechanism) that gave even the most eminent geologists the greatest difficulty of all? No amount of evidence based on animal or fossil distributions, or similarities in sequences of strata, or matching up of mountain ranges across oceans would convince anyone about continents doing this unthinkable thing until a *mechanism* had been found; and I remembered the observation 'To see a thing, you must first believe it possible'.

But a number of plausible mechanisms *were* on offer at the time, one of which – convection currents in the Earth's mantle, driving the continents around like scum on a slowly roiling pan of pea soup – had existed in the literature since 1839. It had been used as a possible explanation for surface features since 1881, in *The Physics of the Earth's Crust*, the first geophysics text ever written, by the Reverend Osmond Fisher (1817–1914). Nor had they been forgotten. Convection currents below the Earth's crust were forcefully advocated through the 1920s by British physicist turned geologist Arthur Holmes (1890–1965). Also, calculations comparing the probable viscosity of mantle rocks accorded well with the speed of the isostatic uplift of Scandinavia after the retreat of the ice sheets, and the rate of movements seen across big faults like the now-famous San Andreas in California. Important work using scaled experimental models (rotating cylinders in tanks full of goo covered with thin skins of wax) was carried out in the 1930s and produced some highly suggestive results.

Indeed, mantle convection is the model geologists still favour today. Yet despite the development of a much-refined model since Holmes's day, *direct* evidence for convection remains elusive. In other words, today 'the mechanism' remains a mystery, but it no longer matters. Because geophysicists have proved with evidence of their own that the continents can and do move sideways, the mysterious mechanism has changed from an insuperable objection to a legitimate subject for research. This means that Charles Ray was right. Many distinguished geologists *had* proposed plausible mechanisms. The acceptance of Wegener's theory did not fail, at least, for lack of convection!

Comparing drift to another geophysical phenomenon, the Earth's magnetic field, helps to show why the 'no mechanism' argument once seemed so powerful. Before the 1940s, when one was finally discovered, the lack of a mechanism for creating the Earth's magnetic field had never dented belief in its existence because scientists could measure it directly. Du Toit's work in Brazil merely provided a better class of the same circumstantial evidence, as did Henno Martin much later when he published on the mirror-image glaciation evidence from Namibia and Brazil. The question was, if the continents were drifting, how could you measure it *directly*?

The scientific world was also moving away from the descriptive, personal methods employed by field geologists like du Toit. America particularly was falling in love with methods that removed the observer from the equation – that 'objectified' observations – rather than demanding the expert view of seasoned experience. Anyone could see the similarity of the opposing coastlines of the Atlantic – just as Ortelius, the first man to see the evidence, had done in his great world map of 1570. He, Alfred Wegener and the mythical 'any schoolboy' could leap to the same conclusion. But if, on the other hand, a *computer* did the matchmaking and confirmed the closeness of the fit, that would be different. And indeed it was, when in 1965 the first computer-matched transatlantic fit was published. The mirror image coasts were no longer dismissible as mere coincidence, a meaningless picture in the clouds. A computer doesn't find patterns where none exist; it doesn't do things by eye or use artistic licence; it applies Euler's Theorem, and doesn't care about the outcome.

Wegener glimpsed this need for direct, objective evidence in a paper he published in German in 1927, and which he quotes in his introduction to the fourth edition of the *Origin*. 'For all that,' he wrote, 'I believe that the final resolution of the problem can only come from *geophysics*.' This conviction was to take him to his death.

Touching the void

Alfred Wegener (1880–1930) was the youngest of five children born to Dr Richard Wegener, an evangelical preacher, and his wife, Anna. Two of his siblings died in childhood and the most notable of the three remaining was undoubtedly his brother and fellow scientist, Kurt. The usual facts you hear about Alfred always include that he and Kurt once held the record for the longest balloon ride, 52.5 hours, and that he later wrote a weighty and respected tome on *The Thermodynamics of the Atmosphere* (1912) in which he came up with the still-accepted mechanism of how raindrops form. He also made many arduous expeditions to Greenland, beginning with a two-year expedition in 1906 when he helped map the island's north-east coast. Crucially, he measured the longitude of Sabine Island and found to his surprise that it appeared to have moved since it had last been measured, about forty years before.

Before the Great War, and shortly after first exposing his heretical drift ideas at a meeting of the Frankfurt Geological Society, Wegener made one more Greenland expedition (1912–13), the first ever to overwinter on the ice sheet, and to do the journey east to west, using ponies. After the war his father-in-law, Vladimir Köppen, retired as Professor of Meteorology in Hamburg, and Alfred got his first proper job as Köppen's replacement.

Two more expeditions to Greenland followed in 1920 and 1930 and in between, Wegener became Professor of Meteorology and Geophysics in the University of Graz, moving Else and their three daughters to the city in 1924. The purpose of both Greenland visits was the same – to establish three scientific stations across the widest part of the icecap.

Bad weather afflicted the 1930 expedition, and in late September Wegener led a desperate attempt to take supplies to the remote central station. The relief party arrived at the end of October; and after a few days spent sitting out some truly dreadful weather, Wegener and an Inuit companion called Rasmus set off for home. No one saw either of them alive again. The following summer Wegener's body was found sewn into his sleeping bag and buried in the ice. He is thought to have died of heart failure. What became of Rasmus, nobody knows.

Wegener had a pressing reason for maintaining his interest in Greenland: his great hope of convincing geophysicists by offering a means of direct measurement of the drift phenomenon, escaping the trap of circumstantial evidence and making drift as real and tangible as the planet's magnetic field.

Wegener reasoned that the best place for seeing measurable drift would be between Greenland and Europe. From the early days he had known about longitude measurements by astronomical methods that had been made by the prestigious Danmark expedition to north-east Greenland in 1906–08, under the leadership of Danish journalist and explorer Ludvig Mylius-Erichsen, who took on the young Wegener as his assistant. Older measurements existed from the general area of this expedition, so Wegener contacted the expedition's mapmaker, J. P. Koch, asking if he could compare their expedition's longitude measurements with them – including measurements taken by the great Irish scientist and explorer Sir Edward Sabine FRS, who went with John Ross's first Arctic expedition in 1818 and made longitude determinations on Sabine Island in 1823. Other measurements not far to the east had also been made, in 1870, by the Germania expedition, and while Koch's measurements were taken in a different location, they could be connected by triangulation.

By combining astronomical and triangulation methods, Wegener and Koch were able to establish a time-series of longitude measurements of Greenland and Europe. These suggested that the gap between them had increased by 4210 metres between 1823 and 1870 (nine metres per year) and by 1190 metres between 1870 and 1907: a massive thirty-two metres per year. The average errors for these measurements were, they thought, in the order of 124–256 metres, which was small enough to make the trend believable.

In the last edition of his great book Wegener also referred to the latest longitude determinations made by the Danish Survey, which in 1922 began a series using 'the far more precise method of radio telegraphy time transmissions'; basically, seeing how long it takes a radio signal to travel between transmitter and receiver. These too seemed to suggest that Greenland was moving west, and by about twenty metres per year.

In fact, none of these measurements was actually demonstrating continental drift. The fact that they appeared to – and drift at such break-neck speeds at that – can only be explained today as a combination of errors, flukes and perhaps also the unwitting interference of those making the measurements. None of the longitude-fixing methods employed at that time was precise enough to detect real drift over periods of tens of years, or even hundreds. But Wegener was not to know that; and his quest for accurate, direct measurement of the process he was convinced broke up Pangaea and created the continents we see today was doomed from the start.

Today you can buy, in minutes and for a few hundred euros or dollars, a navigation system that can pinpoint you or your car anywhere in the world, and tell you how to get from Stoke Newington to Paddington without the need for an *A–Z*. To locate your vehicle to within about ten metres it depends on a system of Global Positioning Satellites (GPS), of which there is a constellation of twenty-four. These all occupy known, fixed positions in orbit 20,200 kilometres up, in a system put in space by the US Department of Defense and completed in 1994. The system allows you, on the ground, to lock on to objects with very precisely known positions, thus enabling it to triangulate on you and pinpoint you precisely. Though GPS was designed and paid for out of military budgets, it was decided to allow its use by civilians after a Korean airliner was shot down in 1983 when it got lost over Soviet territory.

This technology has revolutionized the process of gathering the sorts of information needed to understand the processes of plate motions, all of which require one thing: accurate positioning on the Earth's surface. The plate-tectonic pioneer Professor Tanya Atwater of the University of California at Santa Barbara has written:

> When I began going to sea, our biggest problem was figuring out our position. . . . We were proud if we could locate the ship within a few miles twice a day (by measuring the stars at sunrise and sunset, and then only 'if the weather be good'). The advent of satellite navigation . . . has changed all this. With this system we can now routinely locate the ship to within a few yards every second. When we tell our students about . . . our navigational labours, they look at us as the poor, deprived, primitive ancients.

The sort of GPS kit you might buy for your car has the same sort of accuracy as Wegener's radio-transmission-time method, so you could not use it to detect the widening of the Atlantic reliably over a human lifetime. However, more sophisticated (but still readily available) kit can locate a tripod-mounted detector to within, not metres or even centimetres, but millimetres: easily enough to detect the drift of continents over a few years. But by the time war produced this beautiful technological spin-off, the scientific war over Wegener's theory was long over. The supercontinent of science had re-formed; the resistance of geophysicists, and the opposition of the US geological community, had collapsed under the combined weight of evidence provided by geophysics itself.

As Naomi Oreskes of the University of California, San Diego has pointed out in her seminal study of this great scientific revolution, that geophysical evidence was initially no less 'circumstantial' than the fossils or the rock types of Suess, Wegener or du Toit. However, that evidence was, still, geophysical in nature. These were 'proofs' untainted by association with armchair, qualitative geology; with the opinions of book-learnt experts. They were 'hi-tech' at a time when modernity, equated with computers, automation and machines with flashing lights, was extremely compelling – even to the fogies of traditional geology.

Geologists who had long been convinced of the reality of drift, and who had had the strength of character to regard its mechanism as a problem for geophysicists to sort out, were gladdened by the vindication of their belief. Maybe they were also puzzled that while lack of mechanism had been seen as an 'objection' when all the circumstantial evidence had been geological, now that the circumstantial evidence was geophysical, nobody seemed to worry about it. But they too were also unable to resist the Zeitgeist, and for the most part continued to display that exaggerated and unjustified subservience that geologists have always tended to show to the Queen of Sciences, despite the fact that they (and not physicists) had been right about the age of the Earth and were now being proved right about drift too.

Geologists had seen it in their sediments and palaeontologists known it in their bones: the Earth had to be hundreds of millions of years old, at least. But the great physicist William Thompson (1824–1907), later Baron Kelvin, had insisted otherwise, assuming that the Earth had cooled from a once-molten mass. What Kelvin didn't know about was an alternative source of heat: radioactivity. As is often the case with physics, its objections had been quite correct but only according to what was known at the time. There were more things in heaven and earth than were dreamt of by physics at the end of the nineteenth century. Kelvin, it turned out, didn't know everything (a fact confirmed by some of his other infamous prognostications: for example, that there was 'nothing new' to be discovered in physics, that radio had 'no future' and that heavier-than-air flying machines were 'impossible').

There is absolutely no doubt that during and after the Second World War the advent of geophysics completely revitalized the Earth sciences. But if, as they sometimes do, geophysicists, especially US ones, write or talk as though physics not only proved the reality of drift but even invented it, those who cleave to the geological tradition smile with the same indulgence old men show towards impetuous tyros, and let it drop in the knowledge that *real* history will teach otherwise.

An end of war

When the geophysical evidence finally came, much of it was derived from the ocean basins, where nearly everyone had always thought the answers about continental drift would eventually be found, and where geophysicists such as Tanya Atwater and others eventually found it. Particularly fruitful was a technique that used sensitive ship-borne instruments to map out the magnetization of the ocean bottom. These surveys discovered that the ocean floor is magnetized in stripes created by rocks of either normal or reversed magnetic polarity. In the mid-1960s it came to be realized that this pattern was created when basalt lava, erupting at the mid-ocean ridges where ocean floor is made, became magnetized according to the prevailing magnetic field. Then, as the ocean floor moved away on either side of the spreading centre, new lavas welled up to take their place.

When, as it sometimes does, the Earth's magnetic field flipped and the north magnetic pole sat at the south geographic pole, all subsequent lavas would then be magnetized in the 'reversed' sense – until the field decided (for reasons that are even today not fully understood) to flip back. Ocean floor, which was oldest near the continents and youngest near the mid-ocean ridges, acted like a recording tape, setting in stone the history of magnetic reversals that had happened since the ocean basin began opening, and creating two 'bar code' patterns of ridge-parallel magnetic stripes, one the exact mirror image of the other.

Ship-time is notoriously expensive, and once again it was war that provided the rationale for oceanic magnetic surveying. The reason was simple enough. If you want to detect submarines using magnetometers, you need to see them against a known background. As that research was just beginning, Henno Martin and Hermann Korn would listen to their radio, powered by a truck battery charged by a wind-powered generator, passing the long desert evenings making biltong by their campfire. Martial music and disturbing news from Berlin were a constant reminder of what they were escaping; and long into the night they wondered about the fate of humanity.

It must have seemed odd to Martin to see the drift theory, which he and Korn's researches had long supported, finally receiving its geophysical blessing as a result of research that would never have been carried out had it not been for the very thing that brought humanity its darkest hour. The uncomfortable truth remains that, while science should never be taken as a reason to indulge in it, nothing in human history has done more to improve our understanding of the past and future of our planet than fear of our fellow beings.

8

WRONG-WAY TELESCOPE

We look at him through the wrong end of the long telescope
of Time

D. H. LAWRENCE, 'HUMMING-BIRD'

The naming of parts

Although science is a supercontinent and its citizens participate in a
collective enterprise, it remains a human enterprise, subject to most of
the faults to which humans fall victim. For this reason the desire to
honour heroes is probably as strong among scientists as it is among
generals and admirals; but the nature of the enterprise makes it more
difficult, even when the motives are entirely blameless.

Science mostly honours heroes out of a genuine sense of admira-
tion and respect, and two kinds of scientific advance tend to get
names attached to them. Most are grand hypotheses, but some are
objective discoveries, the equivalent of new mountains or other features of the landscape. These objective discoveries are easier to deal
with (though politically no less fraught). There is no mistaking
Mount Darwin, for example, in Tierra del Fuego, Chile, which
received its name on 12 February 1834 from Captain Fitzroy, the captain of HMS *Beagle*, in honour of the expedition naturalist's
twenty-fifth birthday.

Darwin's theory of evolution by natural selection, however, is often referred to in the twin names of Darwin and Wallace, to give credit to Alfred Russel Wallace, who sketched the same idea (and, crucially, the driving mechanism) in February 1858, the year before Darwin published *The Origin of Species*. Wallace was seized by the same idea as Darwin while recovering from a malarial attack in a beach hut at Dodinga, on the almost unexplored island of Gololo (Halmahera) in the Moluccas. The enforced rest gave him the respite from collecting that he needed for more theoretical thoughts. Darwin, meanwhile, had been most of the time at home in Kent since returning to England from the *Beagle* voyage, wrestling with natural selection, and how to present it to the world, for the best part of two decades.

Scientists usually mean nothing but well by seeking to honour their heroes, yet this will to elect them to the pantheon of the gods so often embroils everyone in acrimonious, futile disputes that an impartial observer may reasonably wonder why the habit persists. As we shall have more to say about the chemical elements later, let us take this example from the world of chemistry, where there is no greater honour than to have one of them named after you.

Elemental forces

In 1997 a long-running and acrimonious dispute over the naming of the element seaborgium (atomic number 106) came to an end. It was the first time an element had been named for a living scientist (Nobel Prize-winner Glenn T. Seaborg, co-discoverer of plutonium). The naming of elements and other chemical substances is the job of a body called the International Union of Pure and Applied Chemistry (IUPAC), of which all national chemical societies around the world are members. Each member state pays its dues according to a complex formula, and the nation that paid most to IUPAC was the USA. It was from this quarter that pressure to name element 106 after Glenn Seaborg, Professor of Chemistry at the University of California at Berkeley, principally came.

The US lobby held that a venerable IUPAC rule banning the naming of elements after living scientists had already been broken in the case of einsteinium (element 99), discovered in 1952 in the debris of the thermonuclear explosion at Eniwetok Atoll. It is probably true that IUPAC would have fallen over itself to name an element after 'the world's greatest scientist', rules or no rules. However, as IUPAC pointed out, the finding was not published until 1955, by which time the great sage had died. So there was no precedent.

Considering the case for 'seaborgium' dismissed, IUPAC proposed that element 106 be named for Ernest Rutherford. Rutherford was the New Zealander who (with others) worked out the structure of the atom and the nature of the various radioactive emissions and who was also the first to realize that radioactive decay could be used to determine the age of the Earth. But the Americans kept up their pressure.

In the end, by virtue of its enormous financial clout, the USA got its way over element 106, which officially became seaborgium in 1997 and Seaborg lived a further two years to savour the crowning triumph of his career.

If naming *things* can cause such problems, it is easy to imagine the difficulties associated with the attribution of *ideas*. Almost every idea has occurred before to someone else, and often long ago, when nobody realized how important it was (like the mapmaker Ortelius and continental drift). To take another example: geology's central doctrine of uniformitarianism, which allows geologists to interpret the past by reference to the processes going on around us today, is usually said to have arisen in the late eighteenth century with the Scots geologist James Hutton. However, it could be said to have been around since at least 55 BC, when the philosopher-poet Lucretius wrote: 'the movement of atoms today is no different from what it was in bygone ages, and always will be. Things that have regularly come into being will continue to come into being in the same manner; they will be and grow and flourish so far as each is allowed by the laws of nature.'

And as for atoms, they go back all the way to Epicurus (341–270 BC), whose school helped lay the intellectual foundations for modern science. If you read original sources, you soon discover that in fact nearly all science's relevant ideas have been there right from the beginning, like jigsaw pieces waiting for someone to see where they fit.

As we shall see, geologists soon came to grasp the idea of supercontinents older than Pangaea, ones that broke up and re-formed, again and again, deep in Earth history. Finding a scientist after whom to name the Supercontinent Cycle shows how hard it is to single out individuals for honours in science's cooperative venture. We shall attempt the choice from three men, in reverse chronological order of their entry into the story: John Tuzo Wilson, John Sutton and John Joly.

New under the sun

All through van der Gracht's volume of proceedings from his New York symposium, Wegener's critics make one point constantly. If 'a Pangaea' really had existed, why did it wait so long before breaking up? Rollin T. Chamberlin of the University of Chicago asked: 'What was happening throughout most of geological time? Why did the continents remain coalesced only to become fragmented very recently?' David White of the US National Research Council agreed: 'How could it happen that conditions favouring the sliding of the continents to the four corners of the earth did not come about until, geologically speaking, almost yesterday?' Joseph Singewald of Johns Hopkins University pointed out: 'The forces called upon by Wegener were operative in pre-Carboniferous time, as in post-Carboniferous time.'

Why then should they only have become effective right at the end of our planet's long life story?

In his lengthy summing-up, the symposium's convener nailed this point right away. There was a logical flaw in the argument, he said. Wegener did not talk about any continental drift that may have happened before Pangaea formed because 'the relevant facts are too little known'. It was not legitimate to infer that continental drift had not operated before Pangaea just because the theory's author had not chosen to address the issue.

Touché . . . But what may have begun as the response of a quick-witted lawyer on behalf of his absent client soon took on the form of a crucial scientific idea, another truly wild surmise. Perhaps super-continents were indeed, like so much else in nature, cyclic. Even in van der Gracht's symposium, one of the very earliest records of a public discussion of drift theory, geologists were hinting at previous phases of continental drift *before* Pangaea. Wegener's book had set out the story of only the most recent episode in a process that perhaps stretched back into the depths of geological time. Drift did not need to be a mysterious one-off, incompatible with uniformitarianism and endlessly repeating histories.

The benign cyclone and the unclassified residuum

Let us return for a moment to the carpenter's bench, and the way the lasagne that was squeezed in the vice to create the mountain range we called the Lasagnides stuck partly to both jaws when the vice was reopened. What would happen next? The answer of course is that new sediments will accumulate in the gap of the open vice (perhaps, on this occasion, a helping of *melanzane parmigiana*), which then become squeezed in turn to create a new range of mountains, the Melanzanides, sitting in roughly the same orientation as its predecessors.

The spatial coincidence of old and new mountain ranges, noted by Swiss geologist Emile Argand as early as 1824, is readily explained in modern, plate-tectonic terms of continental blocks splitting apart again along the sutures that were created when they last collided. And so it was that, as the era of plate tectonics was just dawning, the cyclic nature of continental rifting, oceanic expansion, followed by sub-duction and contraction and finally collision and mountain building, was explained by one of the greatest geologists of the twen-tieth (and perhaps of any) century, the 'benign cyclone', John Tuzo Wilson (1908–93) of the University of Toronto.

In some Huguenot history of geology, pride of place would have to be conceded to Alex du Toit, but 'Tuzo' would run him a close second. He was born in Ottawa to John Wilson, a Scottish engineer, and an adventurous mountaineering woman called Henrietta Tuzo, whose ancestors had crossed the Atlantic and landed in Virginia at about the same time as du Toit's people had been heading south.

Tuzo was one of those charismatic, larger-than-life people whose entry into a room caused heads to turn and conversations to stop. Your eyes went to him; you felt your spirits lifting. His school in Ottawa had made him head boy, and he kept the position for the rest of his life. With his resonant voice he compelled your attention and persuaded you – often against your will – that he was not only right about this but also pretty much right about everything (which, by and large, he was). A positive man, not given to regrets, he would have been brilliant, you felt, at whatever career he had followed, especially, perhaps, politics; and as though to show off his wide-ranging facility, he was also a published expert on antique Chinese porcelain. But global tectonics was his passion, and the plate-tectonic revolution was made for him. It was also very largely made *by* him.

Tuzo had not always been right. Originally a devotee of the con-tracting-Earth hypothesis, he became a convert to drift as he was entering his fifties (by which time he had been Professor of Geophysics at the University of Toronto for a decade). Swiftly recanting his former views, Tuzo saw the way the Earth's mountain belts were often superimposed upon one another, and set about explaining it in terms of plate tectonics. In a classic paper published in *Nature* in 1966 and titled 'Did the Atlantic close and then re-open?' he addressed the coincidence of the modern Atlantic with two mountain ranges called the Caledonides in Europe and the Appalachians in the USA. It was the very first time the new plate tectonics had been extended back to the pre-Pangaean Earth.

These two mountain ranges are really one and the same – except that they are now separated by the Atlantic Ocean, which cut the range in two at a low angle when it opened between them. At one time the two belts had been joined, end to end, Caledonides in the north, Appalachians in the south; and the collision that had created them was one event among many that built the supercontinent Pangaea. Indeed, the matching of the now separated halves of this once-mighty chain provided Wegener with one of his key 'proofs' – part of his geological matching of opposing Atlantic shores.

As van der Gracht pointed out on his behalf, Wegener did not speculate about how his Pangaea had come together. But as the new plate tectonics emerged from studies of the ocean floor and began to revitalize drift theory, the time was ripe to see the break-up of Pangaea as part of a bigger process. Professor Kevin Burke of the University of Houston, Texas, recalls that on 12 April 1968 in Philadelphia, at a meeting titled 'Gondwanaland Revisited' at the Philadelphia Academy of Sciences, Wilson told his audience how a map of the world showed you oceans opening in some places and closing in others. Burke recalls: 'He therefore suggested that, because the ocean basins make up the largest areas on the Earth's surface, it would be appropriate to interpret Earth history in terms of the life cycles of the opening and closing of the ocean basins . . . In effect he said: for times before the present oceans existed, we cannot do plate tectonics. Instead, we must consider the life cycles of the ocean basins.' This key insight had by then already provided Wilson with the answer to an abiding puzzle in the rocks from either side of the modern Atlantic.

Nothing pleased Tuzo more than a grand, overarching framework that made sense of those awkward facts that get thrown aside because they don't fit – ideas that philosopher William James dubbed the 'unclassified residuum'. Geologists had been aware since 1889 that within the rocks forming the Caledonian and Appalachian mountains – that is, rocks dating from the early Cambrian to about the middle Ordovician (from 542 to 470 million years ago) – were fossils that fell into two clearly different groups or 'assemblages'. This was especially true for fossils of those animals that in life never travelled far, but lived fixed to, or grubbing around in, the seabed. By analogy with modern zoology, the two assemblages represented two different faunal realms, just like those first described on the modern Earth by Philip Lutley Sclater and Alfred Russel Wallace.

These two ancient realms were found to broadly parallel the shores of the modern Atlantic Ocean and were described by Charles Doolittle Walcott (1850–1927). Walcott, who had received little formal education, rose to become Director of the US Geological Survey in 1894 and was perhaps one of the most industrious people ever to do and administer science in the United States. He named these assemblages the 'Pacific' and 'Atlantic' provinces; rocks in North America containing the Pacific assemblage, and rocks of the same age in Europe containing the Atlantic.

Had this split been perfect it would have raised no eyebrows among continental fixists because the division would have been easily explained by the present arrangement of continents and oceans. Unfortunately there were some distinctly awkward exceptions to the

rule. In some places in Europe, such as the north of Scotland, geologists found rocks with typical 'American' fossils in them, while in
some places in North America rocks turned up containing typical
European species. In an echo of one of the two scenarios that puzzled
Victorian biogeographers, things were being found close together that
should, by their differences, have been far apart; but with the added
twist that, by and large, they usually *were* far apart.

This conundrum could be explained, Wilson reasoned, if the present Atlantic Ocean was not the first to have separated its opposing
shores: if there had been an older Atlantic, which had closed and
then reopened to form the modern one. According to his idea, the
old Caledonian–Appalachian mountain chain had formed as the vice
shut for the first time, eliminating a now long-vanished ocean that
Wilson called the 'proto-Atlantic'. But when this suture had
reopened, more or less (but not perfectly) along the same line, some
of the rocks squeezed between the forelands had stuck to the opposite jaw of the vice, stranding some American fossils on the European
side and vice versa. The fossil distributions were saying that there had
been continental drift *before* Pangaea. Moreover, if this particular
example could be extended into a general rule, mountain building
itself was inherently cyclic. This process, involving the repeated opening and closing of oceans along ancient lines of suture, has since
come to be known as the Wilson Cycle, a term first used in print in
1974 in a paper by Kevin Burke and the British geologist John
Dewey.

Wilson did not address another interesting problem, which was the
question of exactly *where* on the Earth all this pre-Pangaean action
had played out. From the geological evidence it was clear which continental blocks had done the colliding: which had acted as the jaws of
the vice. But where had they been on the globe at that time?

Wilson did not address this issue because (as van der Gracht might have said) the relevant facts were too little known. They were not long in coming. It soon turned out that Wilson's 'proto-Atlantic' had in fact been sitting right at the bottom of the world. Before 'our' Atlantic had opened, the two jaws of the vice (now represented by North America and Eurasia) had not only opened and closed (and thus helped build Pangaea) but had since migrated north together as far as the Tropic of Cancer before deciding to reopen hundreds of millions of years later, in the great Pangaean split-up.

Wilson's name for this ancient vanished ocean, the 'proto-Atlantic', soon came to seem inappropriate, particularly since the same name was coming to be used for the early stages of the formation of the *modern* Atlantic. Wilson's ocean had been squeezed out of existence by about 400 million years ago: 200 million years before the present Atlantic had even begun to form within Pangaea; so it was no true 'proto-Atlantic' in any real sense. Therefore, in 1972, Wilson's Ocean was renamed Iapetus, which maintains a shadow of the Atlantic link, since in Greek myth Iapetus, son of Earth (Ge) and Heaven (Uranos), was brother to Tethys and Okeanos, and father of the Titan, Atlas.

However, Wilson's great idea was a crucial step forward. It reopened the whole question of 'what happened before Pangaea?' By suggesting that his 'proto-Atlantic' had opened within an earlier supercontinent (just as the modern Atlantic did within Pangaea) he also linked his process to a grander cycle leading from one super-continent Earth to another.

Wilson's originality consisted chiefly of being among the first to consider pre-Pangaean plate tectonics; but it would be stretching things a little to name the whole Supercontinent Cycle after him because his model refers only to 'introversion': the opening and clos-ing of an 'interior ocean', one which opens within a fragmenting supercontinent. And that, as we have seen, is only one way a super-continent can re-form.

Sutton's seed

In 1919 the Sutton's Seeds dynasty was blessed with a son. Unfortunately for this British business, which still flourishes today, it would have to do without the drive and determination of John Sutton (1919–92), who would instead devote his talents to the study of the oldest rocks on Earth. He would eventually join the long line of charismatic leaders of Imperial College London's Royal School of Mines, including its founder, Thomas Henry Huxley, and Sutton's predecessor, Herbert H. Read. He would also marry his near contemporary, Janet Vida Watson (1923–85), to forge perhaps the most formidable husband-and-wife team in geological history.

It was not always easy to work with the great Professor, who suffered from sudden fits of incandescent rage. As his obituarist Professor Dick Selley recalls: 'To be one of his students was like living on the slopes of a volcano. The soil was fertile, the view awe-inspiring, but long periods of productive calm could suddenly be punctuated by an eruption.' Collaborating on almost everything, Watson and Sutton together pioneered the study of the ancient, complex rocks of the Precambrian, but their aim was clearly summed up in the title of a lecture Sutton gave in 1967, 'The extension of the geological record into the Precambrian'. Their aim was to learn how to extend the familiar picture of vanished oceans and the mountain ranges that grew up in their place, back into that mysterious age.

As the great physicist Ernest Rutherford had realized, physics had presented geology with an infallible clock by which to settle the long-standing argument about the absolute age of the Earth. Radioactive elements decay at known rates to products that either themselves decay, or which are stable, in which case the cascade, or 'decay series', comes to an end. The rate of decay is measured in terms of how long it takes a given amount of the radioactive element to be reduced by half, that period being called the element's 'half-life'. Half-lives vary widely in length. The longest-lived atoms of seaborgium, for instance, have a half-life of thirty seconds, while element 104, the one that now bears the name rutherfordium, has a half-life of 3.4 seconds. But these elements, which come into existence for seconds and then just as rapidly decay to something else, cannot exist in nature. Naturally occurring radioactive elements tend to have much longer half-lives, some very long indeed.

To date a piece of rock from its content of a radioactive element, you need to compare the amount of decay product with the amount of the preceding element in the decay series. Then, by knowing the rate of decay (the half-life), you can work out how much time must have elapsed since the rock reached its final form. You have to choose your radioactive element carefully, because, just like clockwork clocks, radioactive ones run down at different rates. You have to choose one that runs for the sort of timespan you wish to measure.

According to Rutherford and his contemporaries, atoms could be thought of as being made up of three basic particles. In their scheme a central nucleus contains positively charged protons with a mass of one and may also contain particles called neutrons with the same mass but no charge. Orbiting the nucleus are a number of negatively charged electrons, whose combined charge normally matches the combined positive charge of the protons. Unlike protons and neutrons, however, electrons have negligible mass.

The number of protons is constant for any element; but elements can contain different quotas of neutrons. As neutrons have mass but no charge, this means that, in nature, some atoms of some elements may differ slightly in weight from others. These forms with different atomic weight are called 'isotopes' of the element because, although different in mass, they all have more or less the same ('iso') chemical properties typical of that element; hence they all occupy the same place ('topos') in the Periodic Table. (Because their weights are different, though, the physical properties of different isotopes are often different, which makes them very useful in geology.) Some isotopes of normally stable elements may also be radioactive: for example, carbon, the element of life.

Carbon exists naturally as three isotopes of differing atomic weight; carbon 12 (the most common), carbon 13 (1.11 per cent of all carbon) and carbon 14 (0.0000000001 per cent). Carbon 14 is radioactive and is continually being formed by cosmic rays bombarding the atmosphere. Neutrons streaming in from space sometimes hit atoms of another element, nitrogen 14, knocking out a proton in the process and creating an atom of carbon 14. As soon as you absorb this carbon 14 – say, when you eat a lettuce that has first absorbed it from the atmosphere – you make that carbon 14 your own. It then begins its slow decay back to stable nitrogen 14 inside your body; but your overall levels of carbon 14 do not change because you top up your levels every time you eat. All food will do this for you, because everything that lives absorbs carbon 14.

But when you die, all the carbon 14 in your remaining flesh and bones goes on radioactively decaying to nitrogen 14; so a test of the carbon 14 in your mortal remains will enable a scientist to determine how long it has been since you ate your last meal. The half-life of carbon 14 is about 5500 years, making it an ideal tool for archaeologists interested in dating once-living things, though these cannot be very much older than about 50,000 years (by which time there's too little carbon 14 left for the technique to work).

Radiometric methods used by geologists to date rock samples are basically the same but depend upon the decay of long-lived elements and their isotopes: substances with decay rates measurable over hundreds of millions, billions or even tens of billions of years. As radiometric dating came to be applied to different rocks all over the world, the first and most dramatic conclusion was that the Earth was definitely not tens of millions of years old, as Lord Kelvin had insisted. Nor indeed was it hundreds of millions of years old, as geologists had suspected. The Earth was *billions* of years old.

Geologists had been more than vindicated; in fact, having grown comfortable with their estimate of 'hundreds of millions of years', they were now presented with a positive embarrassment of time. What is more, nearly all of that embarrassment appeared to fit into the rocks that geologists had until then lumped together in a tiny section at the base of their stratigraphic tables and known (or rather dismissed) as 'Precambrian'. Geologists recognized with some horror that the greater part of Earth history had in fact been written long before complex life had even evolved; that is, before the rocks of the past 542 million years were laid down; which was to say, before those rocks they knew most about even existed. This was a real shock.

The base of the Cambrian Period had been defined according to the earliest appearance of abundant fossils; an evolutionary event caused by the development of hard skeletons that can fossilize readily. Precambrian rocks seemed at that time to be unfossiliferous. There was no reliable way of dividing up these cryptic, complex rocks until radiometric dating came along. And as more Precambrian dates were added to the collection, geologists began to notice a pattern. The dates were not evenly spaced through the 4200-million-year time span. They were clustered.

The older a rock is, the more likely it is to have been buried, cooked up under conditions of extreme heat, pressure or both, partially or completely melted, folded and stretched, and mixed up in the tectonic storm that is mountain building. Every such event will reset the atomic clocks, ticking away within the rock, to zero; so the primary radiometric dates obtained from rocks of this great age do not record the date of their *original* creation, but the date at which they became stable in their present form. In other words, these rocks' radiometric ages refer to the episodes of mountain building in which they have been caught up. The apparent clustering of ages from ancient rocks all over the world, and the broad agreement of these date clusters between different modern continents, soon began to look meaningful. Mountain building, which today we would think of in terms of the collision of tectonic plates, was episodic. And if that periodicity turned out to be regular, which it apparently did, for 'episodic' you could read 'cyclic'.

And so it was, three years *before* Tuzo Wilson published his groundbreaking *Nature* paper on the 'proto-Atlantic', John Sutton published in the same journal a four-page paper titled 'Long-term cycles in the evolution of continents'. In this visionary extrapolation from the global radiometric clustering pattern, Sutton suggested that there was a grand periodicity in mountain-building activity of per-haps 750–1250 million years. His data suggested that 'a structural rhythm of longer duration than the orogenic cycle' might have repeated itself at least four times since the Earth formed.

Sutton termed this the Chelogenic Cycle, because it had been detected in the rocks making up what geologists call the 'shield areas' of the Earth, the ancient kernels of the modern continents. (*Chelos* is ancient Greek for shield, by analogy with the carapace of the tor-toises (*Chelonians* to zoologists) and the defensive posture that Greek soldiers adopted sheltering underneath many shields when under fire.)

One geologist of whom we shall hear more later has already sug-gested, on the strength of this paper, that we should name the Supercontinent Cycle the 'Sutton Cycle'. He is Professor Mark McMenamin, of Mount Holyoke College, Massachusetts, who like Sutton forms one half (with Dianna McMenamin) of a connubial scientific partnership. But this coincidence serves to remind us of a problem peculiar to the older team, which might scupper the chances of the term 'Sutton Cycle' gaining widespread acceptance.

Just as it is hard to see the join between the work of unrelated scientists in the collective activity of science, it is nigh impossible to separate the work of John Sutton from that of his brilliant wife, Janet Watson. Just about everything they did, whether acknowledged in authorship or not, was done together. John's energy, ambition and drive, and Janet's daunting clarity of thought, complemented each other perfectly. Neither would have done the work they did without the other, and there is undoubtedly as much of Janet as of John in the great 1962 paper.

Despite the fact that Janet not only joined her husband as a Fellow of the Royal Society but also became President of the Geological Society of London (something John never achieved), there is a fairly widespread belief that Janet Watson still languishes unjustifiably in the shadow of her powerful husband. Today the gender politics surrounding the scientific legacy of Sutton and Watson is every bit as delicate as that of the Cold War. Naming the Supercontinent Cycle for Sutton alone would not be popular in many quarters; but leaving that aside, the case is a strong one. But there may be an even stronger one.

Like Wegener, Sutton had observed a pattern that cried out for a mechanism. Finding out *what* happened in Earth history is step one in geology; the next step is a search for the *reason* it happened. Why should mountain building be cyclic? Radioactivity, the discovery that gave Earth science both a clock to measure the Earth's age and a mechanism to explain why the planet did not just cool down and die, provided Sutton with the mechanism.

But that idea was neither his, nor Janet Watson's. Moreover, it was the oldest of all, having first appeared in the literature in 1924, waiting for its moment to join, in the right way, with another set of ideas, and finally make new sense.

Trinity – the third man

Trinity College, Dublin, founded by Queen Elizabeth I in 1592, has a tradition of supporting individualistic thinkers. Within its grey granite walls three things came together: long-standing interest in the age of the Earth, the new discovery of radioactivity and John Joly.

One of Trinity's very first graduates was James Ussher (1581–1656), Archbishop of Armagh and Primate of Ireland. Ussher was a versatile scholar, who set himself the task of analysing astronomical cycles, historical accounts and several sources of biblical chronology, to determine the precise date on which his God had created the Earth. His timetable of creation, *Annales Veteris Testamenti*, was first published in 1650; but in 1701 it was incorporated into the authorized Bible, and from that time the Archbishop's calculations came to be seen by believers in much the same dim, religious light.

Although today most people who have heard of Ussher know only about his dating of the Creation to the evening preceding Sunday 23 October 4004 BC, Ussher's project did not rest on the seventh day. After succeeding in his main task the indefatigable Archbishop went on to date other biblical events as well. Adam and Eve, he decided, were driven from Paradise on Monday 10 November that same year, and Noah's ark alighted on the summit of Mount Ararat on 5 May 1491 BC (a Wednesday, apparently).

It is far too easy to laugh at the good Archbishop and his pedantic prose today, not to mention the full English title of his work (1658), which reads: *The Annals of the World Deduced from the Origin of Time, and continued to the beginning of the Emperour Vespasians Reign, and the totall Destruction and Abolition of the Temple and Common-wealth of the Jews. Containing the Historie of the Old and New Testament, with that of the Macchabees. Also the most Memorable Affairs of Asia and Egypt, and the Rise of the Empire of the Roman Caesars, under C. Julius, and Octavianus. Collected from all History, as well Sacred, as*

Prophane, and Methodically digested, by the most Reverend James Ussher, Archbishop of Armagh, and Primate of Ireland.

We do not need, after reading that, to go into the details of Ussher's calculations. Clearly this was a serious scholarly attempt, according to the ruling beliefs of his time, to consult the records of many cultures and answer a nagging question that has only been finally determined by science in the past sixty years.

This question of the age of the Earth was next taken up at Trinity by Samuel Haughton (1821–97), Professor of Geology from 1851, who tried to estimate the Earth's age by adding up thicknesses of sedimentary strata in the belief that their maximum observed thicknesses would turn out to be proportional to the time it took to deposit them. His immensely laborious arithmetic came out with an Earth age of 200 million years, a figure that then seemed so large he scarcely believed it himself. However, his method depended on so many assumptions about rates of deposition in different kinds of rock that you could, by tweaking the sums a bit, obtain almost any answer you wanted. This did not discourage scientists from trying, and Haughton's work was continued by another Trinity professor, William J. Sollas (the eccentric father of Hertha Sollas, who translated Eduard Suess's great book into English). However, it fell to Sollas's successor, the great Irish geophysicist John Joly (1857–1933), at last to make progress in tackling Kelvin on his own terms.

Birth-time of the world

John Joly claimed descent from a line of King's counsellors at the French court dating from as far back as the fifteenth century. His mother went by the title of Julia Anna Maria Georgiana, Comtesse de Lussi. But Joly's father, who died not long after his youngest son was born, lived modestly as a simple country vicar in County Offaly.

Professor John Joly, F.T.C.D.

Contemporary cartoon of John Joly.

Joly was a remarkable all-round intellectual who made important scientific contributions in geology and physics. But, along the way, he also found the time to take first-class honours in modern literature, to invent the first single-plate colour photographic process, to pioneer the use of radium in cancer treatment, devise new navigational techniques and to write poetry, including sonnets on scientific themes, many of which are much better than merely competent. Like du Toit, he rode a motorcycle and also sailed. Joly was a popular man, with his pince-nez, swept-back hair, walrus moustache and rolled r's (an affectation he thought helped to disguise a slight lisp; though many wrongly imagined it was a French accent) and he cut a tall, dapper, even roguish figure among the Trinity dons. He was not without his eccentricities either, notable among which was his habit of wearing a radioactive hat to see if he could detect the effect of gamma rays on his memory.

Joly's was a restless and wide-ranging mind. Like many a don before and since, and despite developing a taste for world travel, he gave Trinity College his life; never marrying, but maintaining an intense long-term friendship with his opposite number in the Department of Botany, Professor Henry Horatio Dixon. Joly worked with the younger scientist on botanical problems, and they are now for ever coupled in the annals of botany for being the first to work out, in 1895, how sap rises. The two men lived close to each other in suburban Dublin, and are today even united in death. Ignoring his friend's wish to be buried in his native Offaly, Dixon had Joly buried in Mount Jerome Cemetery, Dublin, not far from Trinity.

Though a brilliant technical scientist, at his best when solving problems by devising cunning pieces of equipment, Joly seems to have been a little naive. Like many patriotic men of science with some knowledge of the sea, he wrote letters to the Admiralty on the outbreak of war, one of his ideas being to reduce submarine attacks on British merchant ships by building all British ships in the shape of German submarines. However, like Eduard Suess, Joly was no armchair general and was not above taking to the barricades.

On Easter Sunday 1916 Joly, armed with a Lee Enfield rifle, helped to secure his beloved College against the Uprising that was then raging through the city outside. It was a tense time. By Monday, 2000 Nationalists had taken up strategic positions and their leaders had proclaimed an Irish Republic; but the Uprising lasted only a few days before its leaders surrendered. Fifteen of them were executed and up to 3000 more were interned.

Joly put away his rifle, though his Loyalist sympathies remained with him and deceived him badly. As a futurologist he proved no more successful than Lord Kelvin had been, when he predicted that the Nationalists would never succeed in gaining independence from Britain. Only five years later the Irish Free State was established.

Although Trinity first employed Joly as an assistant to the professor of engineering, and then to the professor of natural philosophy, Joly turned increasingly towards geology and used his own colour photographic process (which he patented in 1894) to produce the first colour pictures of minerals in thin section under the microscope. It was following this work, in 1897, that he bid successfully for the vacant Chair of Geology and Mineralogy. He held the job for the rest of his days, and healing the divide between his two main loves – geology and physics – became a lifelong mission. The age of the Earth was too large a question, and too wrapped in Trinity's academic tradition, for him to ignore.

Not thy stars

In 1840 mysterious markings had been discovered on some rocks at Bray Head in County Wicklow: marks evidently made by the feeding activity of some long-vanished organism. The trace fossil was called *Oldhamia*, after the very same Thomas Oldham we have met before in India, but who took up that colonial post with the Indian Survey after serving as Professor of Geology at Trinity.

Joly wrote a sonnet to this humble trace, and Dr Patrick Wyse Jackson, who is today curator of Trinity's Geological Museum and an expert on Joly's life and work, believes that it betrays Joly's special feeling for the immensity of geological time.

> *Is nothing left? Have all things passed thee by?*
> *The stars are not thy stars. The aged hills*
> *Are changed and bowed beneath the ills*
> *Of ice and rain, of river and of sky;*
> *The sea that riseth now in agony*
> *Is not thy sea. The stormy voice that fills*

This gloom with man's remotest sorrow shrills
The mem'ry of thy lost futurity.

Joly, like many a geologist before and since, grew a little giddy staring into the abyss of time, but the uneasy truce over the depth of that abyss finally collapsed when Kelvin revised his estimate of the Earth's age downwards from 100 million to twenty million years. Joly knew there could now be no reconciling Kelvin's conclusions with geologists' gut feeling that the planet simply had too much recorded history to be squeezed into such a short span. Joly attempted to find another way; to search for a different quantitative approach, whose conclusions were not (like Haughton's) open to such differing interpretation, and which could offer a more probing test of Kelvin's conclusions. Independently, he hit on an idea first suggested by Edmond Halley (1656–1742), the first man to predict the return of the comet named after him.

Halley had had different motives from Joly. Although Halley also wanted to expand the amount of time available for geological processes (he was looking for a few thousand years extra), his other objective had been to refute a different, to his mind more dangerous (and much older), idea: namely, that the world was eternal. Halley was not simply trying to burst bonds imposed by Christian dogma but, within that framework, to refute the Greek philosopher Aristotle's idea of an ahistoric, eternal Earth. Aristotle's world without beginning or end, oddly reminiscent of Hutton's nineteenth-century version ('no vestige of a beginning, no prospect of an end') had always offended Christian tradition, because (to use Archbishop Ussher's words) it 'spoileth God of the glory of His creation'.

Just as Kelvin's method was based on two central assumptions – that the Earth was cooling down from an original molten mass, and that no new heat had been added since – Halley's and Joly's idea presumed that the Earth's first ocean had been freshwater, and that all the salt now dissolved in it got there by being washed off the land. If Joly could find out four things – the volume of the ocean, the average concentration of salt in it, the amount of water coming down all the Earth's rivers, and the amount of salt contained in that – then he would have all the information needed to calculate how long it had taken for the rivers to put *all* the salt into the ocean, and thus discover the age of the Earth.

If this sum is done carefully, accounting for all the reverse mechanisms that may return salt to the continents (such as the creation of shallow evaporating seas like the Zechstein) what it actually measures is the average residence time of a sodium atom in the ocean. That is neither an uninteresting nor trivial fact, and the correct average figure is probably around 250 million years, or about the same time that has elapsed since Pangaea began to break up. Unfortunately, it has nothing whatever to do with the age of the Earth. Halley's and Joly's initial assumptions were just as wrong as Kelvin's.

However, Joly did not know this; and when he performed the calculation he came up with an estimate of eighty-nine million years, which was near enough ninety million, which seemed near enough to the 100 million years that geologists had got used to before Kelvin reduced his estimate. When Joly published this research with the Royal Dublin Society in 1899, it was hailed immediately. What was more, geologists (who had chafed under Kelvin's yoke for long enough by this time) at last saw the good Lord being tackled on his own, quantitative, terms – and found it good.

Kelvin's reign was not to last; though, instead of succumbing to attack from without, his chronology collapsed from within. The dawn of the new century brought interesting times for physics, when, with the discovery of radioactivity, subatomic particles and relativity, physicists suddenly realized they actually knew a lot *less* about the world than they had thought. Radioactive decay not only provided the tools to solve the age of the Earth problem once and for all but gave the planet the independent internal source of heat that fatally wounded Kelvin's method and made his hitherto infallible conclusions seem as nonsensical as Archbishop Ussher's. Joly, who corresponded with all the great physicists of his time, was well up to speed with the new thinking. He quickly saw that the Earth could at last be very old indeed. You can sense the excitement in his writing: 'No! The slow exhaustion of primitive heat has not been the history of our planet. Our world is not decrepit by reason of advancing years. Rather we should consider it as rejoicing in the gift of perpetual youth . . .'

And he went on: 'Endlessly rejuvenated, its history begins afresh with each great revolution.'

Joly's halo

Joly's first great insight into the uses of radioactivity concerned something he and other geologists had seen under the microscope, in rocks cut in thin section so that light could pass through and allow all the mineral crystals within to be identified.

For some time it had been observed that crystals of a kind of mica called biotite, which has a beige-brown colour in transmitted light, sometimes seemed to exhibit a rash of dark, circular spots that looked rather like some virus infection on the leaf of a plant. Minerals under the microscope can change their colour when the viewing stage is rotated in polarized light, a phenomenon known as pleochroism. The puzzling spots therefore received the lovely name of pleochroic haloes.

Before the discovery of radioactivity, it was thought that the haloes must result from chemical diffusion of some sort, from whatever lay at the halo's centre, rather like an ink drop on blotting paper. Looking carefully at the haloes, however, Joly realized that this could not be. The haloes were not just fuzzy diffuse blots but were made up of many concentric rings, more like the rings of Saturn. In three dimensions, what presented to the microscopist as two-dimensional rings were, in fact, spheres; and Joly was the first to realize that they formed because a radioactive source at their centre had been sending out high-velocity particles into the surrounding crystal. The colour change was an optical distortion caused by the microscopic damage inflicted by these emanations. What was more, the concentric rings represented different travel times within the surrounding lattice, which meant that different kinds of radiation were being emitted, each with different powers of penetration. From this Joly deduced that it should be possible to work out what the radioactive element was that had given rise to the haloes, since each radioactive element has a distinctive radiation signature.

And so it was that Ireland nearly entered the Periodic Table of the elements, because much of Joly's data defied ready analysis and at one stage he mistakenly thought that he had detected a new element with a hitherto unseen radiation signature. He proposed the patriotic name of Hibernium for this new element; but alas, it turned out that the element in question was already known.

Most pleochroic haloes in biotite are caused by tiny crystals of the mineral zircon enclosed within the mica. Zircon, often used today in jewellery as a substitute for diamond, is chemically zirconium silicate, but within its crystal lattice it is quite common for some zirconium atoms to be replaced by atoms of the naturally occurring radioactive elements uranium and thorium.

Joly, always on the lookout for a new physical measure of the Earth's age, also had the idea that he might be able to use the haloes to date the rocks that contained them. Reviewing various dating methods in 1914 he wrote:

The time required to form a halo could be found if on the one hand we could ascertain the number of alpha rays ejected from the nucleus of the halo in, say, one year, and, on the other, if we determined by experiment just how many alpha rays were required to produce the same amount of colour alteration as we perceive to extend around the nucleus.

The latter estimate is fairly easily and surely made. But to know the number of rays leaving the central particle in unit time we require to know the quantity of radioactive material in the nucleus. This cannot be directly determined. We can only, from known results obtained with larger specimens of just such a mineral substance as composes the nucleus, guess at the amount of uranium which may be present.

Working with Ernest Rutherford, Joly published the results of this method in 1913, using uranium haloes in micas from County Carlow. The research, he wrote, suggested ages of 'from 20 to 400 millions of years'; the halo method, also, was disappointingly vague. But it was, at least, pointing in the right direction. The rock in question was actually about 375 million years old.

But Joly's inventive mind had spotted still more implications of radioactivity for geology. He had realized that more heat was being generated under the continents than was actually being caught in the act of escaping.

In his 1924 Edmond Halley lecture to Oxford University, 'Radioactivity and the surface history of the Earth', Joly told his audience how radiogenic heat had a tendency to build up beneath the insulating cover of the Earth's continental crust. If that went on unchecked, something would have to give. After hundreds of millions of years, Joly believed, rocks deep under the continents would start to melt. The overlying crust would then break up and massive outpourings of lavas – such as are seen in India's Deccan Traps, or the older Siberian Traps, whose eruptions coincided with the break-up of Pangaea – would herald a period of great tectonic instability and fluidity, even rendering possible the outlandish idea of continental drift.

Two years later Joly sat in his study in Trinity College's Museum, writing his contribution for van der Gracht's symposium volume. Though his powers as a scientist were diminishing, he remained open to new ideas and wrote that his theory of recurring cycles of sub-crustal melting now threw the question of continental drift wide open. Offering van der Gracht what support he could, he wrote that the acknowledged fact of radiogenic heat meant that it was now at last 'legitimate to enter upon the problems arising out of continental movement'.

Joly had realized that radiogenic heat, building up inexorably inside the Earth, governed the way the Earth's great heat engine worked. Like all the best scientific ideas, it was incredibly simple. All it needed was the continuous generation of heat, and a thermal blanket to stop it getting out. Van der Gracht noted with satisfaction: 'If radioactive heat does accumulate in the manner discussed, a periodic displacement of the blanketing Sial floats [continents] becomes a requisite . . .'

Today Joly's mechanism is still the accepted basic explanation of why supercontinents break up. A supercontinent sits over the warm Earth like a fur cap sits on your head, holding in heat. However, unlike a fur cap, eventually the supercontinent must break up because the heat has nowhere else to go but up and out. Magmas generated at depth break through the crust and create massive outpourings; convection rising deep below the continent tears it into many smaller landmasses, and sends them off like flotsam in a stream. New oceans begin to open within the supercontinent, whose fragments then, at the speed of your growing toenails, either race all the way around the globe to meet one another on the other side (extroversion), or stall and shrink back on themselves, consuming the young interior oceans in the process as Tuzo Wilson predicted (introversion).

Of this great cycle of the making and breaking of continents, Wilson glimpsed a part. Before him, John Sutton, with his clustered radiometric dates from the Precambrian, outlined the whole. But almost forty years before the plate-tectonic revolution, John Joly already had the explanation of *why* the Earth's grandest pattern operated: by predicting it from basic physics.

The wrong-way telescope of time, which makes the distant seem more distant still, has rendered Joly a sadly diminished figure in our view of history. Yet, if Wegener had discovered a phenomenon looking for a mechanism, Joly discovered a mechanism in search of a process, conceiving an idea that would sleep in the literature until its time was right. Perhaps now that Joly, who also published on possible life on Mars, has had a crater there named for him, the time is right to add to this the honour of the earthly Supercontinent Cycle; for it was his vision that predicted it, and still drives it.

The poet in him certainly rejoiced that the dull fate of gradual heat-death did not, after all, await our beautiful planet. He wrote presciently: 'Our geological age may have been preceded by other ages, every trace of which has perished in the regeneration which has heralded our own . . . a manifestation of the power of the infinitely little over the infinitely great – the unending flow of energy from unstable atoms wrecking the stability of the world.'

Our planet had, Joly saw, an inner life: a life whose warmth demanded a long-term cycle of tectonic activity. Like Halley's comet, supercontinents would keep returning. Mother Earth had a pulse.

9

MOTHERLAND

This film is based on real myths.

NICOLAS CAGE (ATTRIBUTED)

Genesis

In 1934, four years after Wegener met his death on the Greenland icecap and drift theory was perhaps at its lowest ebb, cosmic forces were at work. They were busy causing a set of divinely inspired papers to be translated from the ineffable language of the Universal Father into English, by a complex series of intermediary processes administered by an 'editorial staff of superhuman beings'.

At least, that is the view according to followers of the resulting tome, known as *The Urantia Book*. Like other new religions, its followers make the claim that the teachings contained in its 196 papers are *literally* true. As Harry McMullan III writes in his introduction to *The Urantia Book*, it 'claims to describe reality as it actually is'.

Describing reality as it actually is is, of course, what scientists think they are attempting, though as a rule, if the results involve super-human beings at all, they tend not to set much store by them. Rather, scientists rely on thinking things out for themselves, producing original ideas that explain, as closely as they can manage, the phenomena they observe. As the motto of the world's oldest scientific society, the Royal Society of London, has it, '*Nullius in verba*', or, loosely translated, 'Take nobody's word for it' – and by nobody they really do mean nobody.

It is quite the reverse of the revelatory approach, where in the beginning there always tends to be somebody's Word, which tends always to be with the writer, celestial or otherwise, and with which there can therefore be no argument.

So how spooky must it have been for geology professor Mark McMenamin of Mount Holyoke College, Massachusetts, to discover in 1995 that much of his work to date had apparently been predicted by a sleep-talking mystic from Illinois claiming to be in contact with a Universal Father and his superhuman editorial department?

Mark McMenamin researches rocks from another time, long before the time of Wegener's Pangaea, when all (or most) of the continents were fused into one giant mass. It was also McMenamin who, in notes from 1987, first hit upon a name for it, calling it Rodinia, which he published in a book written in 1990 with his wife Dianna called *The Emergence of Animals – the Cambrian Breakthrough*. By the mid-1990s the name had stuck, particularly (one suspects) because so many of the scientists who work on rocks of this age are Russian. For it derives from the Russian noun *rod*, meaning family or kin; hence the Russian verb *roditz*, which means to give birth to, which in turn gives rise to the noun *rodina*, meaning birthplace, or native land. The McMenamins' Rodinia was the supercontinent around whose shores, and during whose fragmentation, complex life first evolved towards the end of the Precambrian.

Native lands are important to us all, whether we happen to be a French Huguenot living in South Africa or the USA, or a Jewish Viennese born in London. Ultimately the concept is meaningless because, somewhere along the great chain of being, *everyone* has come from somewhere else. But we are all products of the evolution of complex life. Rodinia is the oldest known supercontinent upon whose former existence scientists more or less agree, and so Rodinia can indeed be said to be the birthplace of us all – and of every moving creature upon the Earth.

Urantia

In 1995 Mark McMenamin made an extraordinary fossil find while doing fieldwork in Sonora, Mexico. It turned out to be the oldest known example of a group of enigmatic, long-extinct fossil creatures, which existed before the major divisions of the Animal Kingdom, as we know them today, came into being. He had found the world's oldest Ediacaran fossil.

Nobody really knows what the Ediacarans were, so opinions on the subject among palaeontologists are strong and divided. When I was a student in the 1970s they were known from just a few places world-wide, including Charnwood Forest in the UK and the Ediacara Hills in the Flinders Range of South Australia, after which they were named. But these rare discoveries had occurred in the 1940s and 1950s. Ediacarans were rare, perplexing and, above all, famous. New finds were like hen's teeth.

So there was great excitement in 1977 at Swansea University when one of my lecturers Dr (now Professor) John Cope – a man whose fossil-finding talents are almost supernatural – discovered some new examples of these creatures right on our doorstep just a few miles from the sleepy market town of Carmarthen. The find was also completely unexpected, as it came from rocks that the Geological Survey had long previously mapped as Ordovician, and was made as part of a mapping project that Cope had begun with a group of ama-teurs. Needless to say, as soon as the find's full significance was realized a mechanical excavator was brought in. The whole lot was shipped back to the university for painstaking professional research and a preliminary note of the find to be made in *Nature*.

Ediacaran forms – some palaeontologists feel unable to say for certain whether many of them were animals at all, in the modern sense – display a variety of body plans. To be sure, some may have been the ancestors of later animal groups such as the trilobites; but others seem to show no obvious affinity with any other animal, living or fossil. This was a moment in Earth history when many different forms of life evolved, some highly peculiar when seen alongside modern life, and seemingly showing little or no kinship with anything we would feel comfortable calling either animal or plant.

Many scientists say it is dangerous to assume that all these soft-bodied forms share any common kinship, even among themselves. In fact, so curious are they that some scientists, among them the formidable German geologist Adolf Seilacher of the University of Tübingen, have put forward the view that they represent a completely unrelated evolutionary group that flourished and then vanished leaving no descendants. Professor McMenamin has taken this view further with his highly controversial theory that Ediacarans represent a unique evolutionary creation: in some ways like animals, but also able to grow like plants by absorbing energy from sunlight.

The McMenamin theory suggests that in this early shallow-sea environment surrounding the fragmenting supercontinent Rodinia, these unique life-forms lived, happily sunbathing, fixed to the layer of algal slime and lime mud that then coated practically every square metre of the Earth's shallow sea floor. The McMenamins have called this gentle Eden the Garden of Ediacara: a garden from which these peaceable inhabitants were driven to extinction by two (for them) highly unfortunate biological events. One was the evolution of burrowing (which broke up and destroyed the algal mats on which they sat) and the other was predation. Against these two nemeses the poor

Ediacarans had no defence: they were undermined and grazed out of existence.

When McMenamin got back from his 1995 field season, he enlisted the help of the Mount Holyoke media-relations man Kevin McCaffrey and announced the oldest Ediacaran fossil to the world. To release information in this way is guaranteed to annoy many scientists, who prefer their colleagues to publish their findings in the scientific literature before talking to the media. And sure enough, McMenamin took his fair share of criticism, especially when the story received huge coverage.

But just as his new fossil's fifteen minutes of fame were passing, in October that year McMenamin received a communication from James ('JJ') Johnson, a central figure in the Urantia movement. In the course of his many interviews with the media, McMenamin had described the supercontinent Rodinia as the cradle of complex life; and the unfolding story had begun to ring a bell with Mr Johnson. For there was, he said, a passage in Section 8, paper 57 of *The Urantia Book* that read: '1,000,000,000 years ago is the date of the actual beginning of Urantia history. The planet had attained approximately its present size . . . 800,000,000 years ago witnessed the inauguration of the first great land epoch, the age of increased continental emergence . . . By the end of this period almost one third of the earth's surface consisted of land, all in one continental body.'

These quotations are selective, of course, which is always the key to making the prophecies of mystics look 'uncanny'. If you look at other parts of the same passage from which those quotations come, you can find a rich and colourful mixture of half-correct ideas and plain nonsense. For example: '850,000,000 years ago the first real epoch of the stabilization of the earth's crust began. Most of the heavier metals had settled down toward the center of the globe . . .'

Not bad: the separation of iron and nickel to the Earth's core was indeed an event that took place in the early evolution of our planet, but it happened a lot longer ago than 850 million years. To counterbalance this, as an example of the nonsense among which these little nuggets of correctness lie thinly distributed, we find: 'Meteors falling into the sea accumulated on the ocean bottom . . . Thus the ocean bottom grew increasingly heavy, and added to this was the weight of a body of water at some places ten miles deep . . .'

But the trick of a successful prophet is to say enough things, and to phrase them sufficiently elliptically, so that the occasional correct hits within the general rambling leap out at the prepared mind – just like cloud patterns, or the face of the Man in the Moon. If you are looking for something, in other words, you will tend to find it, which is the very reason why early-twentieth-century American scientists so mistrusted what they saw as the 'selective search after facts' in Wegener's deductive treatise on continental drift. What this story also reveals is that, unlike any other supercontinent that really existed, Rodinia was not envisaged by scientists and later colonized by mystics (like the zoogeographers' idea of Lemuria) but apparently independently 'discovered' by both groups – and it was the mystics who sleepwalked there first.

What happened subsequently to Mr Thompson's communication with Mark McMenamin was 'business as usual'. The devotee was latching on to science because its current conclusions seemed to offer confirmation of a revealed myth. McMenamin, unsurprisingly, fought shy of Mr Johnson's invitation to attend a conference for followers of *The Urantia Book*; but he plainly found the experience thought-provoking, even going so far as to suggest in his book *The Garden of Ediacara* that it might repay scientists' effort to trawl through other mystical maunderings, just in case. It is not possible to be entirely certain how serious he is about this idea. I suspect it might fail simply for lack of volunteers.

What particularly struck McMenamin about the prophecy was that during the mid-1930s – a time when such ideas were distinctly out of fashion – the Urantians had hit upon the existence of a supercontinent dating from one billion years ago (correct), surrounded by a global ocean (obvious, but also correct), at a time when the continents emerged from the ocean more strongly (correct; see Chapter 10) and which subsequently split up about 650 million years ago (about 100 million years out, but still in the right ball park) to form widening ocean basins that became the crucible for the evolution of early complex marine life (also correct). It is also true that until Eldridge Moores and distinguished palaeontologist Jim Valentine wrote their joint paper proposing one in 1970, no legitimate Earth scientist had ever considered the existence of a supercontinent older than Wegener's Pangaea.

Palimpsest

Now that geologists know the age of almost every part of the ocean floor, and can colour it accordingly on ocean-floor maps, it is relatively easy to see how Pangaea fragmented. The ocean floors of the modern Earth are a road map that leads us to Pangaea, by showing us how the modern continents should be put back together. No such map exists for any older supercontinent because the oceans that once opened within them have now all been destroyed, eaten up by subduction and recycled. All that is left of those lost worlds are the broken fragments of ancient continental rock, heavily deformed, embedded within younger rocks, in the shield areas of the world, the ancient hearts of our continents. As the Norwegian geologist Trond Torsvik has written, attempts to reassemble these pre-Pangaean supercontinents 'resemble a jigsaw puzzle, where we must contend with missing and faulty pieces and have misplaced the picture on the box'.

Imagine yourself sailing out of a frozen Baltic port in winter, your ferry butting a channel of black water through the thin ice. As you look over the side at the jagged, jostling floes, you can see a mixture of old and young. Young ice, formed since the last boat passed that way and cleared a lane through the chaos, has been broken for the first time. But that previous boat had itself broken through fresh ice. Pieces dating from that event are still floating about, but are now embedded in floes that tell of two phases of fracturing. Still other floes contain ice fragments of three or more distinct ages, having been through the same process several times, on each occasion the freshly re-broken floes becoming re-frozen into new ice awaiting the passage of yet another ship.

In a similar way the Earth's shields – the ancient hearts of every continent – bear the remaining traces of all the cycles of supercontinent break-up and coalescence since plate tectonics began. During subsequent history many of the pieces may have been destroyed by erosion (Torsvik's missing pieces); but, using the evidence that is left to them, somehow geologists must try to work out which parts of each shield were once fused together in a supercontinent at a given time, and how they fitted together when they are no longer the same shape that they later became. It is one of the most intractable problems in science.

Studying Earth history through interpreting those rocks that have survived is an activity that has a lot in common with the study of ancient texts. Scholars estimate that only 1 per cent of the wisdom of the ancients has found its way to modern times, and the great works of classical antiquity that we have, come to us in the form of documents that were copied, scribbled over and even partially destroyed. Most copies were preserved by pure chance just because of the preciousness, not of the words, but of the material on which they were written.

Take the example of Archimedes and cast your mind back to the principle of isostasy. Although isostasy applies to the way rocks of different density 'float' high or low on the Earth's solid mantle and thus give rise to either ocean or continent, it is really no more than an extension of Archimedes' Principle, which states that any floating body displaces its own mass of the substance in which it is immersed.

Every half-educated person in the world today knows that Archimedes (287–212 BC) shouted 'Eureka!' and leapt out of his bath. But what they should also know is that the story began with a problem put to the great thinker by his patron, King Hiero II. The king was worried that a goldsmith whom he had engaged to make a new crown had adulterated the royal bullion with silver, keeping the remainder for himself. How could Hiero be sure?

Archimedes is reputed to have seen the answer as he lowered himself into his bath, when it dawned on him that every substance has a distinctive density. If you compare the mass of any material with the volume of water it displaces, you have a powerful means of testing its purity, for in the case of gold, any added metal will reduce its density. Having solved the king's problem, Archimedes developed the idea further in one of his greatest works, the *Treatise on Floating Bodies*. However, the only copy of that book to survive to our own day in the original Greek is a rather small, unprepossessing manuscript damaged by mould, fire, and twelfth-century religious zealots. It is called the *Archimedes Palimpsest*, and this precious document came up for auction at Christie's in New York in 1998.

Almost inevitably, there was a legal dispute over its ownership (the Greek Orthodox Patriarchate of Jerusalem contending that it had been stolen from one of its monasteries in the 1920s) but the judge in the case decided against the Patriarchate on 'laches grounds' (that is, because they had left it too long before asserting a legal right). The palimpsest was eventually sold for two million dollars in October 1998 to 'an anonymous buyer from the IT industry'. It is now held by the Walters Art Gallery in Baltimore.

When Archimedes lived and wrote, there were no books like the one you are holding. Archimedes would have copied his theorems and diagrams on to papyrus scrolls, leaving it to succeeding generations to preserve his work by recopying. By the tenth century AD, when what became the palimpsest was originally made in Constantinople, scrolls had given way to more recognizable books composed of leaves of parchment (the preserved skins of sheep, goats and cows) bound between wooden boards. The emperor-scholar Constantine VII Porphyrogenitos and his successors put many scribes to this kind of work, thus rescuing for future generations the rarefied intellectual works of antiquity.

The project was not, however, fully effective, because 200 years later Archimedes' great book was cut up and reused. *Autres temps, autres mœurs*; the great barbarian invasion that was the fourth crusade had sacked Constantinople. In one of the worst disasters ever to befall European culture, many manuscripts were destroyed and the *Archimedes Palimpsest* only survived by chance. The new priority of the age had become the saving of souls; and so Archimedes' text became a *Euchologion*, a prayer book. The new writer took Archimedes' treatise to pieces; scraped off the writing ('palimpsest' is Greek for 'scraped again'), cut out the pages, folded them to half the size, wrote over the original text at right angles and then reassembled the book in its new form. Although this act now seems like desecration, it probably saved the original, since the palimpsest eventually found a home in the Convent of the Holy Grave in modern Istanbul, where it was rediscovered in 1907.

Dr Reviel Netz, Professor of Ancient Science at Stanford University, is a world expert on the works of Archimedes. He has written of the palimpsest: 'A manuscript is a window into the past. It allows us to get a view of a lost world. Some manuscripts provide us with an indirect view only, others with a better picture. What scholars do is to put together all the evidence available, to form a single picture of the past.' Netz could just as well have been describing the work of geologists piecing together the pre-Pangaean supercontinents.

The Earth presents us with the most complex palimpsest of all; it is a text that has been written over, erased, defaced, cooked and reheated; its binding has been broken, its pages lost and shuffled. Each over-writing further obscures everything that has gone before; so that what was originally written may never, indeed, be fully decipherable.

Orogenous zones
Radiometric dating provided geologists with the first clue about the existence of older supercontinents: the clustering of radiometric ages noted by John Sutton throughout the long 'Precambrian' period providing the first hint that mountain ranges, that may today be widely separated from one another, might once have been joined and shared a common origin.

When supercontinents form, all the continental blocks of the Earth come together in a big crunch, eliminating the oceans between them and building mountains in their place as the jaws of the tectonic vice come together. Rocks are pressure-cooked in the roots of each new mountain chain, and radiometric clocks reset. This process is long and complex and does not all happen at once. On the modern Earth, for example, the next supercontinent has already begun to form, following India's collision with Asia and Africa's with Europe. Many other collisions, spread out over the next 250 million years, will take place before the point at which the supercontinent achieves what geologists call 'maximum packing'. But the Earth has a lot of time on her hands. Even dates that fall within 100 or 200 million years of one another, will look clustered within a timespan of (by then) almost five billion years.

Rodinia seems to have formed 1–1.3 billion years ago, as indicated by the clustering of dates. These dates are known as 'stabilization ages' because they mark the point in the mountain-building process when the radiometric clocks were reset. Those rocks that show the joins in this great global collision occur all over the world, but the event itself (called an 'orogeny' because it created mountain ranges) is named for the Grenville Belt of eastern North America. The Grenville Orogeny was what created the supercontinent of Rodinia.

Rocks of this Grenvillian Orogeny are hidden under younger deposits across the eastern and central United States, but crop out in New England, the Blue Ridge Mountains and west Texas. They extend up the great peninsula of Norway and Sweden, as well as down through eastern Mexico, where, once again, they lie mostly hidden under younger rocks laid, like some subsequent historical text, on top. They then skirt the western edge of Amazonia, passing through Bolivia, before diving again beneath younger rocks to the south.

Across today's Atlantic they crop out in Mozambique and Natal, as well as South Africa. In India they are found in the Eastern Ghats. In south-west Australia they are seen in the Darling Belt (which skirts the Yilgarn Craton, the country's richest mineral region, turns the corner at Perth and stretches up the western half of the Great Australian Bight as the Fraser and Albany belts). And because the Bight is the hole out of which East Antarctica was bitten when Gondwanaland broke up, parts of coastal East Antarctica also display rocks whose deformation histories match their Australian counterparts precisely.

So, having found the scars where the supercontinent's component cratons were sutured together into the Rodinian quilt, the next question to be faced is, do they join up, and if so, how? This is much warmer work, and to help them geologists must engage the help of the Earth's changing magnetic field.

Jigsaw

One of the most important geophysical tools to emerge through the 1950s and 1960s involved the discovery that many rocks preserve a trace of the Earth's magnetism as it prevailed when they formed (or became 'stabilized', depending on what kind of rocks you are dealing with). Everyone knows how you can destroy a magnet by heating it, because magnetization depends on the alignment of atoms, which heating disrupts. Rocks are not normally thought of as magnetic, but don't forget that the original 'magnets' that humans discovered were lodestones, natural pieces of the iron mineral magnetite. Many rocks are rich enough in magnetic minerals for them to become very weakly magnetized in the Earth's field.

Rocks that get very hot, such as lavas, or rocks that are cooked deep inside mountain ranges, have to cool down below a certain temperature (called their Curie Point, after Pierre Curie (1859–1906), who discovered the phenomenon) before they can become magnetized. Sediments can also display a weak magnetization, because grains of magnetic minerals will become aligned as they settle out, lending the whole rock a weak magnetic imprint.

When geologists take carefully oriented samples from these rocks and put them into very sensitive magnetometers, they can work out the ancient latitude of the continent – hence where the North Pole was, relative to the continent at the time the rock formed – and the continent's orientation. Because continents drift, their position relative to the magnetic poles changes constantly. Combining palaeomagnetic data with radiometric ages therefore allows scientists to track the movement of a continent over a given time, though by convention they actually do it by pretending that the continent had stayed still and the pole did the wandering.

The resulting Apparent Polar Wander (APW) curves trace out distinctive signatures; and if two continental blocks, which may now be widely separated because of subsequent continental drift, can be shown to have shared APW curves at a certain time, it is a fair assumption that they were once joined together and moved as one – possibly within a supercontinent.

The magnetic field of the Earth can be thought of as a big bar magnet more or less aligned with the planet's axis of rotation. The field's force-lines (which connect the south magnetic pole to the north, describing a giant virtual apple shape in the space around the planet) intersect the Earth's surface at different angles depending on latitude: vertical at the poles and near-horizontal towards the Equator. When you measure the fossil magnetism in rocks, the inclination of the field (as it is known) gives you the rock's original latitude when it formed. The direction of north also enables you to tell the continent's orientation at the time.

But how can you fix the continent's position with respect to longitude at a given moment? Correlation with other pieces in the jigsaw may help. If there are fossils available (as there are after the base of the Cambrian), it may be possible to infer one continent's distance from another at a particular time by the similarity (or lack of it) between assemblages. Obviously, the more similar two assemblages, the more migration was possible between their respective habitats and thus the closer together they were on the Earth's surface. You can also look at sediments. If distinctive mineral grains occur in them, it may be possible to say exactly from where they were eroded. That source may then tie in older rocks that have since become widely separated by later continental drift from the sediments they gave rise to.

A mineral widely used in this sort of study is the humble zircon, the silicate of the element zirconium, which we have encountered already in the nuclei of John Joly's pleochroic haloes. The fact that zircons contain radioactive elements also makes them very suitable subjects for radiometric dating. Zircon grains, eroded from an

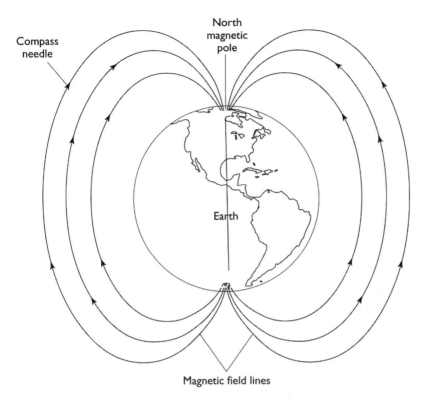

The Earth and its magnetic field. The lines of force of the field describe a shape in space
rather like an apple of which the planet is the core. Note that the angle at which lines
of force intersect the surface of the Earth varies from vertical at the pole to near
horizontal around the Equator. By finding the inclination of fossil remanent
magnetism frozen into rocks, palaeomagnetists (aka 'palaeomagicians')
are able to work out continents' ancient latitudes.

original mass of crystalline rocks, will give the same age as other zircons still *in situ* in the original source rock. You may find sediments
containing zircons with quite different ages, suggesting that they were
derived from elsewhere, plucked from rocks that have since been
rifted off by subsequent drift after the break-up of the original
supercontinent.

Marriage and divorce

Supercontinents, like people, get married and they get divorced. They may also repeat this process more than once. When continents rift, however, they leave traces in the geological record, even after the oceans that the divorcing continents leave in their wake have long been destroyed and forgotten.

Some pieces of evidence, such as the clustered radiometric dates of mountain-building episodes, show us when continents joined together. Different types of evidence have to be used if we are to date the moment when supercontinents broke up. For example, when rocks undergo tension they crack along lines at right angles to the pulling force. Molten magma may rise up and fill these cracks, creating what geologists colourfully term 'dyke swarms' (a dyke being a flat, sheet-like, cross-cutting body of once-molten rock). The dykes' orientation betrays the direction of the tension.

Another consequence of tension is rifting, just as we saw happening in the North Sea Basin at the end of the Permian, when the Atlantic Ocean was beginning to open. In rifting, rock in the valley floor drops down like a keystone in an opening arch, and sediment rushes in to fill the space. Looking at the distribution of rift valley deposits and dyke swarms can help geologists work out how a supercontinent broke apart.

The leading configuration for the supercontinent Rodinia was published by Paul Hoffman in 1991 in the American journal *Science*, and it is based on the assumption that the Grenville-age mountain belts of the world were all created by the elimination of oceans. Hoffman's configuration placed the west coast of North America against East Antarctica, a solution known as SWEAT (South West north America-East AnTarctica), and which was originally proposed by the Scottish-American geologist Ian Dalziel. However, such is the uncertainty in these reconstructions that there is as yet no general agreement among scientists even about this basic configuration. Palaeomagnetic evidence may be geophysics, but it doesn't tell you everything; the results leave a lot of room for interpretation, at least as regards longitude. Also, very ancient rocks lack the fossil controls that can be used to help determine the relative positions of Cambrian and younger continental fragments at given times; and the *geological* controls (matching mountain belts, dyke swarms, rift systems and the like) – as Wegener himself found – need not of themselves compel particular solutions to the jigsaw.

For example, the chief rival configuration to SWEAT puts Australia alongside North America (and is known as AUSWUS, for Australia-Western United States). Other possible configurations are also occasionally dropped into this alphabet soup at international meetings, all serving to demonstrate the extreme difficulty of solving the Rodinia jigsaw puzzle from the scanty evidence of the ultimate palimpsest.

For supercontinents even older than Rodinia the situation is predictably even worse, though just to show that controversy does not necessarily increase proportionately with age, many geologists believe (with Ian Dalziel) that in between Rodinia and Pangaea another supercontinent, Pannotia, was created. In this vision of events present-day Australia, East Antarctica and India rifted off *en masse* from Rodinia about 760 million years ago and became reattached to the eastern side of Africa and Arabia. However, whether Pannotia qualifies as a true supercontinent depends on whether this event did any more than build the megacontinent Gondwanaland. Opinion on this remains resolutely divided. One recent textbook on the subject, for example, makes no mention at all of Pannotia among the panoply of pre-Pangaean supercontinents.

We have seen how supercontinents may form by two processes, for which geologists have borrowed the psychological terms 'introversion' and 'extroversion'. Introversion is another name for the Wilson

Cycle, sometimes also called 'accordion tectonics', whereby a continent rifts apart, forms an ocean within itself and then closes again along the same line, destroying the interior ocean and forming a new range of mountains more or less where an older range once stood. Extroversion simply envisages this rifting continuing, so that the original supercontinent is turned inside out and all its fragments meet one another along their leading edges somewhere else on the planet.

Tales from topographic oceans

The solution to the question of continental drift did indeed lie at the bottom of the ocean, as many geologists suspected. The problem that geologists have in putting pre-Pangaea supercontinents like Rodinia back together (in other words, in distinguishing between such possible solutions as SWEAT and AUSWUS) is the lack of ocean floor from that time, because it has all long since been destroyed. However, it would be an immense help if, even without that 'road map', they could somehow tell whether a given supercontinent (whose existence and approximate date of fusion we should be able to tell from such evidence as clustered radiometric ages) formed by introversion or extroversion. The question is, how?

Consider this. In the case of an interior ocean (like the modern Atlantic, opening between fragments of a disintegrating Pangaea) all the ocean floor that has formed is obviously younger than the break-up of Pangaea. If the two sides of the Atlantic should decide to close again and form Chris Scotese's Pangea Ultima, the ages of all ocean-floor rock that will have to be consumed will fall (at their oldest) between the dates of Pangaea's initial break-up and (at their youngest) its eventual reunification.

But on the other hand if Roy Livermore's vision comes true and

250 million years from now his Novopangaea forms by the opposite process of extroversion, the ocean floor that is consumed in forming the new supercontinent (mainly the Pacific Ocean) will all have lain outside Pangaea at its state of maximum packing: the floor of an 'exterior' ocean called Panthalassa. Nearly all of this ocean floor therefore formed before Pangaea was fully assembled; and therefore nearly every piece of it, especially the very first pieces to be destroyed, would yield ages *older* than the break-up of that supercontinent.

In other words, in introversion, all ocean floor consumed in making a supercontinent will be *younger* than the break-up of the previous supercontinent, while in extroversion it will be *older*.

This presents a potential method of telling which of the two mechanisms broke apart and then formed supercontinents older than Pangaea. But how useful could it be? It relies, after all, on dating ocean floor that is consumed in the creation of a new supercontinent. But surely, you may ask, subduction *consumes* all ocean floor, so it is no longer available for sampling. If there is indeed none left, how can this idea move us forward?

The answer lies in remembering the difference between perfect models and imperfect reality. It may be true in textbook diagrams that subduction destroys all ocean floor, but it is not so in real life. Real subduction is not the clear-cut business represented in these tectonic cartoons; and sometimes, instead of diving down into the Earth like they should, something goes wrong and pieces of ocean crust become scraped off on to the continents to become parts of mountains: true 'topographic oceans'.

Geologists have long recognized these distinctive rocks. They consist of three basic elements: basalts, erupted underwater and thus forming characteristic pillow shapes; the vertical dykes that fed these submarine eruptions with lava; and the glassy mineral chert (silica, or silicon dioxide) sitting between the pillows. The pillows typically

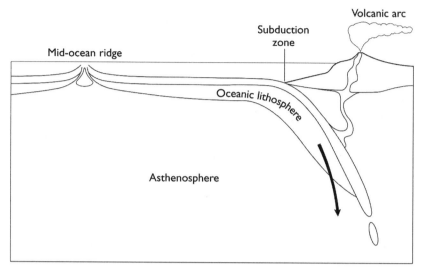

Cartoon representing plate tectonics. Ocean floor, produced at the mid-ocean ridge, is pulled back down into the crust at subduction zones. However, the process is not this neat in nature, and bits of the ocean floor get scraped off to survive on top of the continental plate, as 'ophiolite'.

have chilled margins (small crystals, or even glass) where the hot magma met the ocean and cooled very quickly. Below, the dykes that fed their eruption formed as tension at the mid-ocean spreading ridge opened up long, parallel cracks at right angles to the direction of tension. The cherty sediments between the pillows, rich in silica, partly precipitated from solution and partly derived from the skeletons of such creatures as sponges and the microscopic diatom. (There are few other microfossils because, in the low temperatures of the deep sea, those with skeletons made of calcium carbonate dissolve away.)

Pillow lavas, dykes and cherts form a classic threesome first noted in 1905 by Gustav Steinmann (1856–1929) and grandly named the 'Steinmann Trinity' in his honour by the eccentric Scots geologist Sir Edward Battersby Bailey.

But the Steinmann Trinity is only part of what geologists now call an 'ophiolite suite', an ocean-floor remnant that may run to thicknesses of three to five kilometres. Deeper still within the sequence, below the sheeted dykes, come massive, coarsely crystalline rocks called gabbros. These rocks are the solidified remains of the magma chambers that fed the dykes and that cooled more slowly because of their greater volume and depth, and thus formed rock with the same chemistry but larger crystals. Lying below the gabbros are the deepest rocks of all, including peridotite (which has sometimes been altered by seawater to form a well-known rock called serpentine, often used for ornaments, ashtrays and cheeseboards). These dark-green rocks are slices of the Earth's mantle.

Only with the coming of plate tectonics were these distinctive rock sequences recognized for what they are: the last remnants of long-vanished ocean crust, scraped off on to the continent but destroyed everywhere else by a subduction process that once drew two continental crust blocks together in a mountain-building episode. If enough ophiolite formed during the accretion of a particular supercontinent could be dated, the spread of results should enable us to tell if that supercontinent formed by introversion or extroversion.

However, there is a big problem with this idea, and it has to do with resetting of radiometric clocks. Ophiolites commonly have three important 'ages'. There is the date they were created, their 'magmatic age'. Their second radiometric age is an overprint that dates from the point at which they began to be involved in subduction processes, as the increase in heat and pressure began to alter their constituent minerals. Then there is their third age, which is when they were scraped off on to the oncoming continental crust. For this method to work, geologists need to know the rock's first, true age: the date when it was first born from the mantle.

Pannotia

In 1991 Paul Hoffman wrote a paper with the title 'Did the breakup of Rodinia turn Gondwanaland inside out?' According to this model of how Rodinia fragmented, about 760 million years ago a megacontinental landmass made up of the continents we now know as Australia and Antarctica rifted off from Rodinia along a line that now defines the western edge of North America. (This is not the present-day coast of North America, because since Pangaea split up, the USA and Canada have ridden over much of the ancient Pacific Ocean floor, colliding with many small landmasses on the way. These have accreted as what geologists call *terranes* to the west coast, and built up the mountainous western seaboard of North America in a process akin to the way that litter collects against the top of an 'up' escalator, when the steps are finally subducted into the bowels of the machine.)

In making this move, the Australia-Antarctica continent opened up an interior ocean that became the ancestral Pacific. Ancient ocean floor encircling the fragmenting Rodinia was subducted, and the process continued until Australia-Antarctica had swung round and collided with another continent consisting of South America-Africa, which at that time were still joined along the line that would one day open to form the present South Atlantic. Australia-Antarctica thus became fused with South America-Africa, creating a megacontinent we have seen before: Suess's Gondwanaland. Many geologists also believe (with Ian Dalziel) that at this time other pieces of continental crust were also very close together, possibly fused, and have given this supercontinent assemblage the name 'Pannotia' ('all southern continents').

Because the ocean that was consumed in this process was all 'exterior' to Rodinia, it provides a known example of extroversion; and a possible test for the dating method, because the maximum ages obtained from any ophiolite remnants of the consumed ocean floor should pre-date the break-up of Rodinia.

When Pannotia subsequently split, about 550 million years ago, the interior oceans created by this event, such as Iapetus, Tuzo Wilson's so-called 'proto-Atlantic' separating present-day North America from Western Europe, were subsequently destroyed as the next supercontinent (Pangaea) was created. They were, we know, destroyed by the accordion tectonics of the Wilson Cycle process – that is, by introversion. Dating fragments of *those* vanished ocean floors should therefore yield ages *younger* than Pannotia's break-up.

If a large enough sample of true ages is gathered from ophiolites preserved during the formation of Pannotia, they should fall in a period entirely before the date of Rodinia's break-up because the ocean floor they represent was all exterior to that supercontinent. Conversely, if the same is done for oceans that formed between the break-up of Pannotia and the formation of Pangaea, the ages obtained should fall within that interval of time because all the ocean floor they represent formed as an interior ocean. If these predictions hold up, our method should show that Rodinia extroverted to form Pannotia and Pannotia introverted to form Pangaea.

Geologists Professor Brendan Murphy of the Tectonics Research Laboratory at St Francis Xavier University, Nova Scotia, and Professor R. Damian Nance of Ohio University have pursued this technique with great success. Their elegant joint research has, since 1985, resulted in a much clearer picture of how supercontinents assemble. Murphy and Nance have looked at rocks associated with the assembly of Pannotia (about 600 million years old) from the Borborema Belt of Brazil, and the Trans-Saharan and Mozambique Belts of North and East Africa. Rodinia began to split apart about 760 million years ago. So, if Pannotia formed by the consumption of exterior ocean surrounding Rodinia, the formation dates should come out at between 760 million years and about 1100 million years.

Taking on the mantle

But I still haven't answered the main question: just how, exactly, do you find out such dates reliably? Simple radiometric dating, as we have already seen, allows you to find out when the atomic clock was last reset. But rocks from the floors of vanished oceans, now anomalously preserved in the mountain belts that replaced those oceans when they closed, have all been involved in mountain building and had their clocks reset. Simple radiometric dating would reveal the date the mountains formed, but that is not what we want. We want to get at the time these ocean floors formed at a mid-ocean ridge. We want the birthday, not the date of the funeral, or the mid-life facelift. We want to know the very first time those ocean-crust rocks were created by volcanic melts, erupting at a mid-ocean ridge; the very moment they were derived from the mantle and became part of the crust.

To find out that crucial birth-time, Murphy and Nance have developed a method that combines radiometric-dating techniques with the tendency of isotopes of elements to separate out: become differentiated during natural processes because of their different atomic weights. The technique is complex and beautiful, and it involves using isotopes of two unusual elements: samarium and its daughter element neodymium.

These two elements are very similar in many ways. Both sit close together in Group 3 of the Periodic Table, which occupies two long lines at the bottom of the chart, and which are collectively known as the 'rare Earth elements'. On the top row (the 'Lanthanide series') you will find neodymium (Nd) at number 60 and samarium (Sm) at number 62. The atoms of these two elements are about the same size, and they react similarly in chemical processes because in their cloud of electrons both elements have the same number available for forming bonds – two. However, something different happens to neodymium and samarium atoms when rocks in the Earth's mantle begin to melt and form the magma that will eventually build the ocean floor.

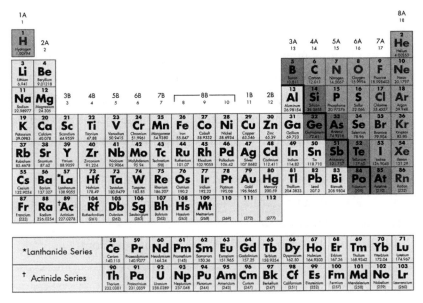

The Periodic Table of the Elements. The Lanthanide Series is one of the so-called 'transition element' groups (horizontal lines) arranged in a double line below the main diagram.

Lanthanide-series elements display a very strange property. As their atomic number (and hence their atomic weight) rises, the atoms actually *shrink* rather than expand. This is known as the Lanthanide contraction. When melting begins, rock becomes a mixture of liquid and solid. Lanthanide contraction makes the heavier and denser Lanthanides (such as samarium) more likely to stay where they are in the solid phase, while the lighter, bigger atoms such as neodymium will tend to prefer to melt. This means that neodymium atoms have a higher tendency to leave the mantle, and samarium to stay put. Mantle rocks have therefore become progressively depleted in neodymium since the planet formed 4700 million years ago.

If you could count all the atoms of samarium in the Earth and divide that number by the number of atoms of neodymium, you would get a figure that expressed their relative global abundance. If the two elements were present in equal amounts, the number would, of course, be one. If you found more neodymium than samarium, then the resulting number would go down. For example, if there were twice as much neodymium as samarium, you would be dividing two into one, and the result would be 0.5. These numbers are called ratios, and they are useful in expressing the relative abundance of two things.

If you do the calculation for real, taking the Earth as a whole and dividing the number of Nd atoms into the number of Sm atoms, it works out at about 0.32: the 'bulk Earth' ratio. Or, putting it another way, there are about a third as many samarium atoms as neodymium atoms inside planet Earth.

But because, through geological time, Nd has been continually leaving the mantle in volcanic melts that have gone to form oceanic crust, Nd is more concentrated in the crust than in the Earth as a whole, so the samarium-to-neodymium (Sm/Nd) ratio of crustal rocks is lower, at 0.2. Conversely, Nd has been depleted from the mantle, so the mantle now contains less than average amounts of that element, making the Sm/Nd ratio of 'depleted mantle' higher (about 0.5, in fact).

So far we have not considered the matter of different *isotopes* of samarium and neodymium, but have thought of all the isotopes of each element collectively. However, both samarium and neodymium have isotopes aplenty. Samarium has seven natural ones, one of which, samarium-147, is unstable and undergoes radioactive decay to neodymium-143. This radioactive transformation is very slow indeed, with a half-life of 106 billion years – or over seven times the age of the universe. Neodymium has nine isotopes, seven of which are stable. One of them, neodymium-144, is not a product of any radioactive decay series, so it does not change in concentration in a given rock with time and therefore can be used as a benchmark.

The gradual decay of samarium-147 (^{147}Sm) to neodymium-143 (^{143}Nd) therefore has the effect of making ^{143}Nd more common through geological time in all rocks, gradually increasing the 'bulk Earth' ratio between its daughter element ^{143}Nd and the unchanging ^{144}Nd. However, remember that there is more samarium in the depleted mantle than in the crust because of neodymium's tendency to fractionate into melts that head upwards. Therefore the increase through time in the ratio of the two isotopes of neodymium will be faster in the depleted mantle (where the parent element samarium is relatively abundant) and slower in the crust, where Sm is rarer.

This means that the isotopic signature provided by the ratio of ^{143}Nd to ^{144}Nd gives the rock a fingerprint for its place of origin. Because radioactive decay processes are known, unchanging and predictable, you can then, by extrapolating backward, determine when the rock from which you obtained the sample left the mantle. It is as though the rock has never lost its accent: you can take the melt out of the mantle, but you can't take the mantle out of the melt. Moreover, because samarium and neodymium are almost identical chemically, this fingerprint is almost indelible: it is almost unaffected by most subsequent changes that a rock might undergo.

This allows geologists to take more or less any rock that formed by the crystallization of magma – even if it formed by the remobilization of previous crustal rocks – and work out when its chemistry began to go its own way and depart from the isotope chemistry of the depleted mantle. This works because, in the end, nearly *all* rocks were originally derived from the upper mantle. As long as there has not been contamination from other melts with different histories (and this is usually evident from the field geology) the method is a sound way of telling when the rocks were first born.

This crucial date, the rock's first birthday, is called the Depleted Mantle Model Age, abbreviated as T_{DM}. This is the tool we have been looking for: a way of telling when these pieces of ocean floor, now preserved in mountain belts, first left the mantle at a mid-ocean ridge.

Back now to the rocks. By taking as many samples as possible from ophiolite suite rocks that were emplaced during the elimination of the ocean we are studying, and then comparing their T_{DM} ages with the date that the mountain range formed (which we know independently from straightforward radiometric dating of that event), we should at last be able to determine whether that process was one of extroversion or introversion.

Using this technique, Murphy and Nance found that the rocks from Brazil that were caught up in the formation of Pannotia after what would become West Gondwana rifted off the previous supercontinent Rodinia, provide T_{DM} ages of about 1200 million years, and that similar rocks from south-west Algeria and southern Morocco come out at between 1200 and 950 million years. In the Mozambique Belt the T_{DM} ages come out at between 800 and 900 million years. From this they concluded that the vanished ocean, whose tombstone is those ancient mountain belts, formed part of Rodinia's exterior ocean, because the T_{DM} ages are all older than the date of the break-up of Rodinia. As predicted by the Paul Hoffman model, Pannotia formed by the extroversion of Rodinia. Rodinia turned inside out.

In the case of Pangaea, the supercontinent after Pannotia, the results have been, as expected, very different, but also consistent with predictions. The main mountain ranges that formed as Pangaea reached its maximum packing – the great suture scars that mark the healing up of oceans – are the Appalachians in the USA, the Caledonian mountains of the UK and Norway, the Variscan mountains of southern Europe and the Urals of Russia. We know that Pangaea's predecessor Pannotia began to fragment at about 550 million years, when oceans like Iapetus began to form. So, as oceanic rocks caught up in the formation of Pangaea by the destruction of those 'interior' oceans, they should all have T_{DM} ages of less than 550 million years. No data are available as yet from the Urals, but data from the other mountain belts all indicate that they were derived from the Earth's mantle less than 550 million years ago. Pangaea formed by the accordion tectonics of introversion.

Quadrille

So, even from times 1000 million years in the past, geologists now are finding ways of determining how continental fragments moved about the face of the planet, consuming the ocean before them as they went. But this stately dance of the continents, which, like partners in a quadrille, move apart, twist around one another and come together again in new combinations, has not gone on for ever. Geologists can recognize the probable existence of even older supercontinents than Rodinia, though they will remain even more dubious and controversial until new techniques can be found to recover more information from the geological record.

But what makes ancient secrets more difficult still to unlock is the probability that in Earth's deepest past the tectonic processes that operated were quite different from those we see around us today. For these distant pasts, the present no longer provides the key. The Earth may then have worked in ways perhaps as different from the tectonics of today as the Ediacara garden's inhabitants may have been from modern life forms. And when it comes to life on Earth, geologists tracing the evolution of the planet from its Hadean origins are increasingly wondering how biology itself may owe both its beginning, and its drive to complexity, to the workings of our planet's inner life – as told in the greatest palimpsest of all.

10

BIRTH

We must be humble. We are so easily baffled by appearances
And do not realise that these stones are one with the stars.
HUGH MACDIARMID, *ON A RAISED BEACH*

When geologists hit upon the notion of constraining their dreams of the past in terms of processes observed operating today, it made geology a science. Although debate has raged ever since about whether those processes always operated with the same *intensity* at all times in the long history of the Earth, the method still held together. But it only held for those youngest, lightly scraped and overwritten pages of the great palimpsest that was open to geologists of the nineteenth century and much of the twentieth: namely, the 542 million years since the beginning of the Cambrian Period, when the age of animals (and eventually plants) with easily fossilizable bodies dawned, heralding complex life's long march – or random walk – through increasing complexity.

But with radiometric dating came the shocking realization that this segment only represented about the last 12 per cent of Earth history; and that, in the conventional stratigraphic tables of the time, the tiny unregarded plinth of complex rocks labelled 'Precambrian' on which the geological column rested, actually contained nearly all the time that had elapsed since the planet formed. It was like digging a well,

GENERAL TABLE OF THE STRATIFIED SYSTEMS
AND FORMATIONS, Etc.

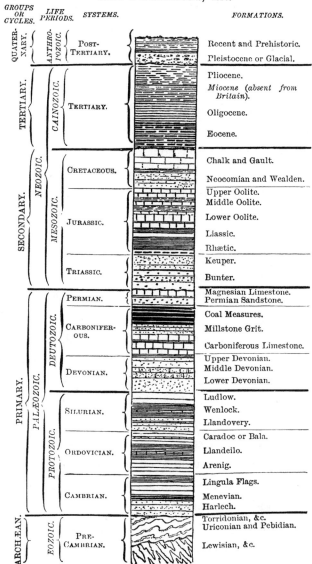

GROUPS OR CYCLES.	LIFE PERIODS.	SYSTEMS.	FORMATIONS.
QUATER-NARY.	ANTHRO-POZOIC.	Post-Tertiary.	Recent and Prehistoric.
			Pleistocene or Glacial.
TERTIARY.	CAINOZOIC.	Tertiary.	Pliocene.
			Miocene (absent from Britain).
			Oligocene.
			Eocene.
SECONDARY.	NEOZOIC. / MESOZOIC.	Cretaceous.	Chalk and Gault.
			Neocomian and Wealden.
		Jurassic.	Upper Oolite.
			Middle Oolite.
			Lower Oolite.
			Liassic.
			Rhætic.
		Triassic.	Keuper.
			Bunter.
PRIMARY.	PALÆOZOIC. / DEUTOZOIC.	Permian.	Magnesian Limestone. Permian Sandstone.
		Carbonifer-ous.	Coal Measures.
			Millstone Grit.
			Carboniferous Limestone.
		Devonian.	Upper Devonian. Middle Devonian. Lower Devonian.
	PROTOZOIC.	Silurian.	Ludlow.
			Wenlock.
			Llandovery.
		Ordovician.	Caradoc or Bala.
			Llandeilo.
			Arenig.
		Cambrian.	Lingula Flags.
			Menevian.
			Harlech.
ARCHÆAN.	EOZOIC.	Pre-Cambrian.	Torridonian, &c. Uriconian and Pebidian.
			Lewisian, &c.

An old-fashioned stratigraphic table, dating from 1898. The basal part labelled 'Precambrian' actually contains 88% of Earth history. Taken from Charles Lapworth's *Intermediate Text-Book of Geology* (Blackwood & Sons). From the collection of the author.

only to find what you had thought to be solid bedrock giving way into a black, bottomless, unsuspected cavern, loud with a vast and terrifying silence.

As geologists now looked for ways to decipher the rare and often badly damaged pages of the Precambrian chapter, they began to realize something else deeply shocking. They began to see that there were things in the deepest places of Earth history for the unlocking of whose secrets the present no longer provided the key. True, up to a point the old tools still worked; after all, a poorly sorted conglomerate full of mud and cobbles and boulders the size of a man were still probably dumped by glaciers. (However, the same difficulties and controversies would attend their interpretation in the late twentieth century as in the mid-nineteenth, the only difference being that now the arguments were more sophisticated.)

But the problem went deeper than just interpreting the meaning of particular rock types. The Precambrian world that the old tools and other new tools such as isotope analysis revealed was not the familiar, endlessly cycling Huttonian or Lyellian world, ringing to what Thomas Hardy described as 'the full-fugued song of the universe unending': a world with no beginning, offering no prospect of an end. By contrast the vast spans of Precambrian time were dominated by progressive, secular processes that had wrought permanent change upon the evolving Earth system.

The more geologists thought about it, the more reasonable this began to seem. Just like the life of a human being, Earth's growing years left their indelible marks upon her; and yet despite her difficult early traumas, by middle age she was leading a much more stable, routine, almost (but not quite) *predictable* life. She had reached a time in her life in which it was almost impossible to conceive of things ever being that radically different. Indeed, if things were so radically different then from now, perhaps they were *too* different for geologists to build a scientific picture of the Precambrian world? If the rule of the present could no longer be used to measure out the ancient world, what price the scientific method? How could a geologist's imaginings of these very different times be constrained?

But all was not lost. Geologists turned first to the immutable laws of physics and chemistry; and in addition they found something new: the emerging techniques of computer modelling. Using these new approaches, John Sutton and Janet Watson's dream of opening up the Precambrian gradually came to be realized. It was the final confirmation that the uniformitarian visions of Lyell and Hutton did not, and could not, do full justice to Earth's chequered past. Moreover, as the idea that human activity might be affecting the Earth System became familiar to followers of current affairs, the whole question of how the climate works (a question rooted in how it evolved into its current state since the birth-time of the Earth) lent Precambrian geology a sudden relevance, even urgency. The purest of pure science, this expedition to an alien planet whose curiosity-driven mission directive had been drawn up without the slightest idea of practical application, suddenly moved politically centre-stage. For the tale of the Precambrian has proved to be a litany of terrible climate disasters, all of them brought about – or at least hastened – by life itself, and the Supercontinent Cycle.

Genesis

Erasmus Darwin (1731–1802), Charles's eccentric, versifying, visionary ancestor, in his epic poem *The Temple of Nature* wrote: 'Organic Life began beneath the waves.' In 1871 his grandson, on the other hand, would speculate, in a letter to the Director of Kew Gardens, the explorer-botanist Sir Joseph Hooker:

It is often said that all the conditions for the first production of a living organism are now present, which could ever have been present. But if (and oh! what a big if!) we could conceive in some warm little pond, with all sorts of ammonia and phosphoric salts, lights, heat, electricity, etc. present, that a protein compound was chemically formed ready to undergo still more complex changes, at the present day such matter would be instantly devoured or absorbed, which would not have been the case before living creatures were formed.

So who was right? Did life start in the seas, or in a little lightning-struck dish of lukewarm primeval soup? Today's leading theory about how life came to planet Earth suggests that the older Darwin, on this occasion, came nearer the mark. Life, it now seems likely, originated not in a superficial pond, but deep below the waves on the gloomy floor of the Earth's early oceans. What is more, it did so long before even the continents were fully formed and set sail across the globe. It isn't only Hugh MacDiarmid's stones that are 'one with the stars': life is, too.

The Earth began to accrete from a disc of space debris around 4.7 billion years ago, in a hellish birth-time colourfully referred to by some geologists as the 'Hadean' eon, though this name is not recognized by the International Commission on Stratigraphy, the body that decides such things. It prefers instead the rather more prosaic name 'Eoarchean' for the earliest, bottomless section of the Earth's life story: the true and final plinth of the modern stratigraphic column below which there was simply no Earth.

As it vacuumed space debris orbiting the young Sun, the Earth gradually heated up. Gravitational energy from incoming bolides was converted into thermal energy. Deep within, the iron and nickel in the mixture separated out from the molten rocky materials and sank into the planet's core, where it still remains, an eternal but querulous dynamo driving (and occasionally flipping) the Earth's magnetic field. Up above, and for perhaps as long as 500 million years, our planet was a cratered, volcanic spaceball, sporadically molten, dark, sterile; blasted by solar wind, flayed by ultraviolet light, too hot for oceans, too hostile for life.

Although Lord Kelvin was quite wrong about how old the Earth could be because he assumed it had just cooled by radiating heat into space from its original molten state, it is true that our planet was a very much hotter place four billion years ago. This was partly due to bombardment and gravitational heating; but it was also due to the much greater abundance at that time of highly radioactive isotopes.

Remember that all radioactive decay series eventually end up in stable isotopes, or at least in longer-lived and much less radioactive ones. Shortly after the solar system formed and the rocky planets coalesced from space junk, Earth's nuclear reactor burned much hotter than today, just as radioactive waste, which will eventually become harmless, is at its hottest when it is fresh.

Because the young planet needed to dissipate maybe five times as much internal heat, the mantle convection systems deep below its crust would have been smaller and more active than today's. Therefore, with greater production of volcanic material at surface, and faster movements among smaller plates, the number of spreading and subduction zones would have been greater. Moreover, the crust that they formed (and consumed) would also have been much thicker: somewhere between twenty-five and fifty kilometres. This has led geologists like Eldridge Moores to ask when this non-uniformitarian change from thick to thin oceanic crust took place, what the environmental implications of that change would have been, and whether it happened gradually, or more suddenly.

Birth of the continents

The crust of the newly accreted Earth would have been everywhere of the same composition (roughly speaking) as modern ocean floor, simply because this is the basic stuff of the Earth. Lighter rocks, which float high as the continents of today, had to be distilled from that crude material by the fractionation of lighter elements. Thus the silica and aluminium compounds, identified as 'SiAl' by Eduard Suess, had to be separated from the heavy ones, made predominantly out of silica and magnesium ('SiMa'). So the continents cannot always have been the same size as today: they had to grow. The oceans, too, may well have been more voluminous than today because the hotter mantle could contain less water, chemically locked up in minerals, than it can today. Some scientists think there may have been twice as much water at surface, making the early Earth a truly pan-thalassic water-world.

An example of how the process of continent formation might have started can be seen today on Iceland. By processes of partial melting and melt extraction in the system of conduits under the island, magmas of granitic (classic SiAl) composition are being formed. This is why, although Iceland is always portrayed as a black, basaltic place befitting its position on a mid-ocean ridge, in fact up to 10 per cent of its rocks are of light, granitic type.

Iceland sits astride a hot, hyperactive stretch of the Mid-Atlantic Ridge, and has been forming for a mere sixteen million years or so. Just like the early crust of the Earth, because the amount of volcanic material erupted there is higher than average, Icelandic crust is twice as thick as normal ocean crust, which all helps the lighter 'continental' type rocks to separate out as the cooling magma circulates in the plumbing below. Because of their lower density, sialic rocks – once created – would then remain at the surface, gradually coalescing as they jostled and fused by continuous minor collisions forming the *protocontinents*.

As the Earth cooled further, oceans began to condense, hydro-logical cycles of evaporation and precipitation began to operate and the erosion and deposition of sedimentary rocks could really begin. The earliest evidence of erosion comes from rocks over four billion years old, in the form of those amazingly robust and persistent micro-scopic mineral grains that John Joly saw at the centre of his pleochroic haloes: zircons.

As we have seen, when rocks undergo melting, and elements of dif-fering atomic weight separate out between solid and liquid phases – a process called fractionation – different isotopes of the elements (despite their identical chemical properties) behave differently accord-ing to their slightly different weights and measures. Some prefer life in the melting pot, while others tend to remain in the solid. If conti-nental crust is continually fractionating from the Earth's primitive material, isotope ratios within the different rocks generated will grad-ually come to differ from average, or 'bulk Earth', values. Thus, even in one single, tiny grain of zircon, distinctive fractionated isotope ratios remain as the telltale signature of early crustal processes that generated suites of continental rocks of which those tiny crystals are today the only surviving remnants. Truly, a whole world in a grain of sand.

Mark Harrison at the Australian National University in Canberra and his co-workers have been studying zircons that were eroded, re-incorporated and then sealed within younger rocks about 4.4 billion years ago. These rocks come from the Jack Hills in Western Australia, and themselves constitute one of the oldest pieces of continental crust on the planet. Harrison and his colleagues have tested these grains for two isotopes of the element hafnium (Hf).

The zircons contain very low concentrations of another element, lutetium, whose radioactive isotope ^{176}Lu decays to hafnium; so the researchers believe that the ratios of ^{176}Hf to ^{177}Hf which they find in these grains are very close to the primitive ratios that prevailed when they formed – that is, at the original zircon-containing rock's absolute age, determined independently using uranium-lead dating methods. Those ratios are characteristic of fractionated sialic crust. What this suggests is that continents were not only forming, but even being eroded and their detritus redeposited, within as little as 200 million years of the Earth's accretion: that is, between 4.4 and 4.5 billion years ago.

In fact, the more geologists look into this question, the more it seems as though fractionation processes had produced continents of almost modern dimensions fairly soon in Earth history. There is little direct evidence of this because very little continental crust dating from the Archean (between 4.5 and 2.5 billion years ago) survives today. Barely 7 per cent of rocks in the present continents are that old; all the rest have now long been eroded and recycled as younger continental crust. So how is it possible to tell how big Archean continents were? The answer comes in the form of meteorites.

The stony meteorites are the leftover raw material out of which the early planets were formed. This means that the early Earth originally displayed the same overall concentration of uranium as meteorites still do. In geological processes, uranium tends to fractionate into continental crust; the more uranium there is in the continents, the less there will be in the mantle. Therefore, one measure of the total amount of continental crust in existence at any given time is the degree to which mantle rocks are depleted in that element. The surprising result of testing those few surviving mantle-derived rocks from the late Archean (2.4 billion years ago) is that they appear to be just as uranium-poor as modern mantle.

Some time in the Archean (probably earlier rather than later) the continents became 'fully grown'. Continents may since have fused and split, danced and skated over the globe like the faces of a Rubik's cube in maddening and almost untrackable ways; but their total volume has remained about the same for the greater part of Earth history, the product of a dynamic balance between the erosion of existing continental crust and the production of new.

And into a very dark corner of this hostile, water-enveloped world, deep below the turbulent surface of the first global ocean, life was born. Earth was a young, hot planet; spinning so rapidly on its axis that a day lasted no more than five violent, storm-racked hours. Its acid, anoxic ocean was raised in frequent massive tides by a Moon that had not yet wound away to its present distance, though it may well not have been visible in the feeble light filtering through the murk of gas and dust thrown up by meteorite impacts that would, from time to time, vaporize sea and rock alike.

Slumbering green

Julian Huxley FRS (1887–1975), son of the great Thomas Henry Huxley, in a wistful poem to a dancer whose performance had captivated him, lamented that he felt

> *Weary of plodding science, where the vision*
> *Must for achievement clothe itself in clay . . .*

Nowhere do you find the vision obscured more often, or more thickly, than in the dreaded conference abstract: a tight little knot of compressed and overwrought jargon with which scientists announce early findings to a room of critical colleagues they are hoping to impress. But they are not all like that.

'Earth agglomerates and heats. Volcanoes evolve carbon oxides, methane and pyrophosphate. Convection cells, stacked in the planetary interior, begin the cooling process. An acidulous Hadean ocean condenses from the carbon oxide sky. Stratospheric smogs absorb a proportion of the Sun's rays. The now cool ocean leaks into the crust and convects . . .'

The author of this was Professor Mike Russell, then a research fellow at the Scottish Universities Environmental Research Centre in Glasgow and currently 'distinguished visiting scientist' at NASA's Jet Propulsion Laboratory in Pasadena, California. Russell is a world-renowned expert on the early Earth and the planetary geology that can give rise to life. Together with his colleague Allan Hall, he has written widely about how life may have originated; but in order to do that, any scientist has to ask what sounds like an unanswerable question: namely, what exactly *is* life? For Russell and Hall, the best way to approach this loaded question is to stick to what can be observed and measured. Instead of asking what life *is*, Russell asks what life *does*. And to put it simply, life exists to make a mess.

'A living cell assimilates nutrients, uses energy and generates waste. It consists mainly of carbon-based ("organic") molecules that also contain hydrogen and other elements. Their defining structural feature is a mainly waterproof container, the cell membrane. Inside is a watery solution with a high concentration of organic molecules as well as some inorganic salts.'

To tackle this question of what life 'does', scientists need to understand natural sources of energy and what forms of energy are involved in life processes. Says Russell: 'What does a waterfall do? It drains water from a greater to a lesser height, giving the water a lower *gravitational* energy. What does a warm spring do? It is a plumbing system draining *thermal* energy from underground and dissipating it on the surface. So, what does life do? Life is a chemical system that drains and dissipates *chemical* energy. For example, animals and plants gain chemical energy from sugar in food by burning it in inhaled oxygen, a process we call respiration.'

On Earth today, nearly all life ultimately derives its energy from the Sun, which drives the whole process. Green plants catch the rays and use them to ensnare carbon from the air and water from the soil to produce big organic molecules, with which they build their tissues. The waste product of this process (called photosynthesis) is oxygen, which animals then use to break down the carbon compounds in the food they eat, thus releasing their energy and generating the raw materials they need to grow. Plants are the origin of nearly all life as we know it because only they can use pure energy to build their bodies; bodies that animals at the bottom of the food chain must eat to build theirs. All flesh, as the Bible says, is grass. This has been the way the living world has worked for billions of years. But not always. Certainly not from the beginning.

Mike Russell and Allan Hall think that the first living cells formed on the floor of the Earth's first newly condensed oceans, where warm, alkaline submarine springs focused chemical energy, and the mixing of the hot spring water and seawater caused simple chemicals to precipitate out as thin, inorganic films.

Under today's oceans, at regions where the heat-flow is high, such as the mid-ocean ridges, seawater leaking into the hot rocks of the seabed is superheated (sometimes as high as 700°C), becomes charged with minerals and is then extruded into the cold sea at submarine hydrothermal vents known as 'black smokers', because the dissolved minerals immediately come out of solution with the sudden fall in temperature and create the impression of chimney smoke. But these waters can be hot enough to melt lead, and are only kept from boiling by the intense pressure. They are far too hot for organic life to have originated in or near them, even though they are often densely colonized today by specially adapted organisms.

However, a similar hydrothermal process can also happen far away from the hot ridge, on much older (and cooler) ocean floor. At these vents, percolating seawater itself is responsible for creating heat, by hydrating the mineral olivine (the basic mineral component of the mantle) to create another mineral, serpentine. These springs reach only relatively moderate temperatures (the hottest being about 170°C); but these off-ridge alkaline vents (first discovered in 2001, though predicted by Professor Russell some years before) do grow much larger than their hotter equivalents on ridges. They are also a distinctive ghostly white (since they are mainly made of calcium carbonate) and the largest of them are known to rear up over thirty metres from the ocean floor.

Ocean-floor springs also contain many essential minerals which all organisms require, such as phosphorus, nickel and zinc. At a time when the Earth's surface was inimical to the existence of organic molecules, here was somewhere safe and protected, where life could form uninterrupted. The gradients of temperature and acidity/alkalinity could provide the energy while the minerals brought chemical food within a solid structure built of a mix of carbonates, silica, clays and sulphides of iron and nickel. Mike Russell, who began his career apprenticed as a chemical engineer studying nickel catalysts, recognizes in this system what an industrial chemist would call a 'continuously regulated flow reactor'.

A further hint that life truly originated in these dark, submarine places soon after the oceans first condensed perhaps 4.4 billion years ago, is that the microbes still living today among modern hot springs include some of the most genetically simple life-forms on Earth. Vent communities may form a closed ecosystem, but Russell believes life eventually escaped from them to colonize the world. And that great escape was probably brought about by the processes of plate tectonics: familiar to us as elements in the Supercontinent Cycle.

Russell's insight into the way life originated on the Hadean ocean floor began while he was entertaining his son Andy by making 'chemical gardens' with the sort of chemical kit you can buy from science museum shops. These crystal gardens seem to 'grow' in a plant-like way; but the snaking, knobbly tubes rising from the beaker bottom are purely inorganic, forming at the interface of the crystals that the experimenter drops into the solution. However, Russell remembers that on the night after starting the chemical garden, Andy started to break it up. 'Deaf to my pleas to join us at supper,' Russell remembers, 'he announced, from behind the locked bathroom door, "Hey Dad, these things are hollow!"'

Suddenly Russell thought about the puzzling patterns he had seen in the rocks of an ancient ore body he had studied in Ireland, long after he had left the chemical industry and qualified as a geologist. He had seen columns, chimneys and bubbles in the rock, all made of iron sulphide. These had once been natural 'chemical gardens', a garden, he soon realized, in which the seeds of life could have grown. As he discussed the idea the next day with his colleague Allan Hall, he saw how organic molecules could have become trapped and concentrated along those flimsy chemical membranes precipitated as the hot alkaline spring water, rich in minerals, met the cold, acidic sea. These small pockets of proto-cells could have encouraged more complex and unlikely reactions to take place, just as every life-form today uses a membrane to protect and concentrate its contents. Russell says: 'The precipitation of chemicals on mixing of solutions forms a barrier, preventing further mixing and precipitation. At the warm spring we envisage the formation of a special precipitate that not only formed a barrier, but also provided a template for the assembly of chains of organic molecules, and acted as a catalyst for electrochemical reactions.'

He thinks that along such thin chemical boundaries organic molecules like amino acids, the building blocks of proteins, first became concentrated. These organic molecules would have formed a little deeper in the columnar 'flow reactor', where water and its dissolved chemicals were reacting with iron and nickel-rich minerals. The precipitate membrane would then capture and concentrate other chemicals that could participate in reactions, he thinks; but eventually this system would evolve, by a process of 'organic take-over', into a cell membrane consisting entirely of organic molecules. Russell and Hall also believe that by acting as a template, the iron-sulphide precipitate could bond chemically to, and assemble a sequence of, the molecular components of RNA, a chain molecule very similar to DNA, which plays a supporting role in genetic evolution.

'Once organized on the iron sulphide, this RNA could influence the assembly of amino acids into proteins, as well as the assembly of further chains of RNA, and, finally, of DNA. Eventually, these new large organic molecules could reproduce themselves through the interaction of DNA, RNA and proteins, without any need for the original iron-sulphide template.'

But how could life, assuming that it was indeed born on such ocean-floor vents, have escaped from its abyssal refuge? Russell thinks that during life's first few hundred million years the only safe escape route would have been *down*. Early organisms could have migrated into the ocean floor, with its warm underlying sediments, and subsisted there on a diet of trapped hydrogen and carbon dioxide. This was the beginning of the so-called 'deep biosphere', the mass of microbes that sits silently and invisibly in the pore spaces of the Earth's lithosphere, but whose total mass even today outweighs all the living things around us on the planet's surface.

There, safe from lethal solar rays, early life played its long waiting game; until the rocks, now teeming with endolithic microbes, completed their plate-tectonic journey and became involved in a subduction zone. While most of the sea floor slid down into the mantle, some would inevitably become scraped off on the overriding plate to form an ophiolite suite; and, thrust up into the shallows, a few bacteria would have found themselves deep enough to be protected from harmful rays but shallow enough to use the Sun's longer-wavelength light to build organic molecules from carbon dioxide. It was the beginning of photosynthesis.

Rust world

Earth's early acid-bath oceans contained huge amounts of dissolved iron, which by mid-Archean time had been spewing into it from ocean-floor hydrothermal springs for a billion years or more. This iron was held in solution in both its positively charged ionic forms: ferrous iron (an atom lacking two electrons and thus having a positive charge of two) and ferric iron (lacking three electrons, and with a positive charge of three). The ferrous iron mostly came into the water from very hot springs, whereas the ferric form would have been created by oxidation of this ferrous iron by the weak sunlight, or by the waste product of the earliest forms of photosynthesis.

The novel condition of matter we call life probably originated some time around 4.4 billion years ago, probably around the time that the Australian rocks containing those robust little zircons, with their tell-tale hafnium isotope ratios, were forming. As the mantle roiled beneath, small blocks of light continental crust were forming at the Earth's surface and sticking together, growing like plaques of scum on a lake: a process that geologists have called cratonization. By three billion years ago these had coalesced into the first known recognizable supercontinent: Ur.

It is interesting that cratonization did not occur at the same pace all over the Earth's surface. At the time that the supercontinent of Ur was forming (it connected areas of South Africa, Madagascar, Sri Lanka, India, Western Australia and parts of East Antarctica), elsewhere on the Earth small convection cells were still producing 'greenstone belts', the typical product of the Earth's earliest tectonics: small kernels of continental rocks surrounded by highly deformed, ocean-floor-type rocks.

Yet on Ur, processes of erosion and deposition were taking place in modern style. Between about 3.1 and 2.7 billion years ago, for example, on what is now the western side of southern Africa, large thicknesses of sediments were laid down in the Witwatersrand and Pongola basins. Here, geologists have been able to decipher distinctive environments, such as intertidal flats and braided streams (some containing gold), sand ripples, mud cracks (and the very first glacial deposits to form on Earth).

Ur was also, it seems, the first supercontinent to experience glaciation, which in turn supports the idea that even then the global climate was not too dissimilar from that of today. That said, the atmosphere contained very low concentrations of oxygen, perhaps less than 1 per cent of modern levels. How glaciations could happen at all when the atmosphere was essentially full of greenhouse gases is one of those questions that further research into the Earth system must answer. As the Supercontinent Cycle that would continue to our own day was just beginning, emergent life unleashed the first of its many great climatic catastrophes on Earth, as oxygen – this new and toxic gas, hitherto just the rare product of chemical dissociation – began to appear in rising quantities as life escaped from the upper deep.

Life's ancient sleep had ended. It came to the surface as bugs living in the pore spaces of rocks that, instead of being subducted beneath them, were scraped off on to continental masses. Under those shelf-seawaters, lit by the sickly light of a weak Sun perhaps 20 per cent less luminous than today, beneath the suffocating twilight of a greenhouse atmosphere full of carbon dioxide, methane and nitrogen, life learnt to photosynthesize.

Oxygen is a highly reactive gaseous element that now makes up about 16 per cent of the atmosphere. The growth of oxygen has not been steady: there have been times when much more oxygen was present in the air than now; for example during the Carboniferous Period, just as Pangaea was forming. Coal forests then covered much of the planet, pumping out oxygen and sequestering carbon. These forests were therefore even more inflammable than they are now, and the higher oxygen levels also made it possible for dragonflies of the time to have wingspans of a metre.

But oxygen was almost entirely absent from the (to us) toxic atmosphere of the early Earth, composed as it was mainly of volcanic exhalations. That this was so is proved by the fact that Archean sediments often contain detrital grains of the mineral pyrite (iron sulphide), which means that this unstable chemical could then survive intact at surface. You may have found Jurassic fossils preserved in this mineral (also known as fool's gold because of its brassy colour) on the beach at, for example, the world-famous fossil site of Lyme Regis in Dorset, England. It formed there under anoxic conditions below the Jurassic seabed. But if you have ever taken these fossils home you may also know how short a time they last, all shiny and lovely, unless protected from air. For oxygen corrodes the pyrite and one day you find that your fossil has disintegrated into an ash-like powder. Such is the fate of all pyrite at the Earth's surface today – but not in the Archean. Yet as soon as life had escaped from its submarine lair and gone green, this was to change.

Free oxygen was pouring into a world where elements such as iron and sulphur had always been able to exist quite happily – on land or in the ocean – in their unoxidized (or 'reduced') form. Today unoxidized iron cannot exist on the Earth's surface for long, and we see the evidence all around us in the rusting hulks of old motor vehicles. But before 'free' oxygen could build up in the air to levels that could support oxygen-breathing animals, the products of the first photo-synthesizing organisms had to oxidize all the iron and sulphur. These elements constituted a vast oxygen 'sink'; one that geological evidence suggests was not completely filled for as long as 1.6 billion years.

The 'rusting' of the Earth began about 3.5 billion years ago, which is how we can date the approximate emergence of photosynthesis. The first of a new and distinctive rock type began to be deposited in the oceans: rocks that are today the world's dominant source of iron ore. They occur on every continent and contain about 100,000,000,000,000 (one hundred trillion) tonnes of iron, nearly all of it deposited between 3.5 and 1.5 billion years ago. These spectacular rocks, marking the second-biggest biochemical event since the creation of life, are the Banded Iron Formations, or BIFs.

BIFs consist of interbedded iron ore and chert, a flinty rock that began as a gelatinous deposit formed by the chemical precipitation of volcanic silica from seawater. As fast as the first photosynthesizers pumped their waste into the air, reduced iron dissolved in the acidic seas combined with the reactive new element and precipitated out as insoluble oxides in unimaginable quantities: not only in the shallow shelf seas, but presumably all over the oceans too. This means that the 100 trillion tonnes of iron that has survived to the present day is but a tiny fraction of the total iron deposited at this time, partly because the ironstones that formed on the deep ocean floor must have been destroyed by subduction. David Dobson and John Brodholt of University College London, in research first published in 2005, think they know exactly where this ancient iron now is: deep inside the planet, sitting at the boundary between the core and the mantle.

Seismologists have long puzzled over the nature of what they call Ultra Low Velocity Zones (ULVZs) between one and ten kilometres in thickness, where seismic waves, generated by earthquakes far above, suddenly and anomalously slow down as they travel close to the core-mantle boundary. If BIFs did form everywhere in the Archean ocean, including over ocean floor, because a BIF is 25 per cent denser than the mantle, once subducted it would tend to go on sinking until it hit – literally – rock bottom, where the silicate mantle touches the roiling, molten iron and nickel of the outer core. If Dobson and Brodholt are right, the processes of life, far from being a superficial veneer, have affected this planet to its deepest interior.

For this reason, it may also be that life changed the speed of the Earth's rotation. The sinking of such a huge mass of iron to the core-mantle boundary would have had an effect similar to the gradual drawing in of a pirouetting skater's arms, causing the days to get shorter. Scientists remain undecided about this idea, however, since we know (from the growth lines on fossil corals) that since the Devonian period (a mere 360 million years ago) the drag of the ocean tides has tended to *slow* the Earth's rotation. However, the period over which the BIFs were sinking occurred so long before the evolution of complex life that its opposite effect on day-length remains a plausible but untested idea.

Oxygen's chemical sinks began to be near 'full' about 2.3 billion years ago. By 1.9 billion years ago, BIFs had ceased to form (though they did make a brief comeback, as we shall see) and levels of oxygen in the atmosphere began climbing towards modern levels, with further disastrous consequences.

During this time more large continental groupings were beginning to coalesce. At about 2.5 billion years a second more northerly supercontinent emerged, called Arctica because the Arctic Ocean opened through its middle, incorporating much of northern and central Canada, Greenland and Siberia. At about two billion years a second northerly supercontinent, called Baltica (consisting of much of northwest Europe as far as the modern Urals), assembled and eventually collided with Arctica to form (1.6 billion years ago) another larger continent, known variously as Nena, Nuna or Columbia.

Also at about two billion years ago, a southerly megacontinent dubbed 'Atlantica' (because eventually the Atlantic Ocean would open through it when Gondwanaland broke up) united much of north-eastern South America with West Africa and the Congo, and forming the heart of what would one day be West Gondwana. These three components of different ages, ancient Ur, Nena/Nuna/Columbia and Atlantica, finally came together about one billion years ago to form the first true supercontinent, motherland of complex life, Rodinia.

But this is to race ahead. After BIFs ceased to be deposited and oxygen began to be more freely available, sediments exposed on land became oxidized too, and 'red beds' became common. At about 1.7 billion years ago there was also a marked increase in weathering, because thick deposits of quartzite, a rock type made from pure silica sand, suddenly became widespread all over the continents. Such an outbreak of silicate weathering would have had a drastic cooling effect on climate, as carbon dioxide in the air became combined with rock materials to form bicarbonates, which were washed into the seas. What happened then, effectively sequestering the carbon and keeping it from returning to the atmosphere, relied on another development happening in the oceans.

One by one, the panes of the greenhouse that had kept Earth warm for billions of years were being smashed.

Lasagne world

It is one of the most remarkable discoveries in deep time that for most of the history of the biosphere, a period four times longer than all the time since the first complex fossils began to be preserved, the most advanced life-form on the face of the Earth, and the absolute pinnacle of evolution, was *slime*.

When a slimy surface, coated in some kind of alga or bacterium, is exposed to sand and mud on the floor of the sea, the sediment tends to stick. The organisms then grow through the sediment to expose a fresh surface of slime to the sea, and in this way great thicknesses of thinly layered rock consisting of interleaved slime and lime can be deposited. These lasagne-like sedimentary structures are called stromatolites.

The very oldest stromatolites come from 3.5-billion-year-old rocks in Western Australia, a time when we still find no firm fossil evidence for the existence of photosynthetic organisms. But 'stromatolite' just means 'layered stone', irrespective of what creatures built it. It is entirely possible for many different sorts of bacteria to produce layered structures, and what exactly made these very ancient examples is still a mystery. However, at around 2.7 or 2.8 billion years ago something changed dramatically. Isotopes of carbon obtained from rocks of this age indicate the activity of a group of primitive organisms called Archaebacteria, and highly resistant organic molecules called steranes also suggest the presence of another bacterial grouping, the cyanobacteria: photosynthetic bugs formerly known as 'blue-green algae'.

Before levels of atmospheric oxygen rose very far, these cyanobacteria had the new trick of photosynthesis more or less to themselves and began to coat the shallow shelf seas surrounding the early supercontinents of Ur and Arctica. Their heyday was brief, though, and came to an end about 2.2 billion years ago, when oxygen in the environment built up enough for oxides of nitrogen (the nitrates) to form. Nitrates are a powerful plant fertilizer which cyanobacteria can do without; but from that point stromatolites built by invigorated true algae (primitive plants whose cells' genetic material is organized into a nucleus) took over. Lasagne World was born just as Rust World was dying.

Algal stromatolites were soon coating every available square kilometre of shelf sea, and building the first massive limestones in the geological record. In fact, by about one billion years ago (at just about the time that all the continental fragments of the Earth were coming together in the supercontinent Rodinia) limestones were forming a greater percentage of sedimentary rocks than ever before or since. Some of these limestones are of truly awesome dimensions, kilometres thick, and the effect of trapping all that calcium carbonate – coming on the heels of increased continental weathering – was to reduce the amount of carbon dioxide in the atmosphere even further.

Earth's atmosphere has two main 'greenhouse' gases that cause it to trap the heat of the Sun: methane and carbon dioxide. Methane, also known as natural gas, is by far the more powerful; but it is very susceptible to oxidation (burning, by another name). With global oxygen building up, free methane in the atmosphere went into steep decline. Now, as carbon dioxide started to be drawn down by weathering on the emergent continents and then locked away in limestones at the bottom of shelf seas, the planet began finally to lose the last threads of its insulating atmospheric blanket.

Stromatolites reached the high point of their distribution just as Rodinia was forming: a major turning point in Earth history. For it seems from geological evidence that at about this time, one billion years ago, a fundamental and irreversible change occurred to the way our planet functioned.

Freeboard

The fact that the modern ocean basins are more or less the right size to accommodate all the water currently available to fill them is one of the great apparent coincidences of geology. True, from time to time, notably when supercontinents break up and volcanic mid-ocean ridge systems become more active, the ocean basins become a little less voluminous as the ridges all over the world swell up. (Imagine yourself lying in a full bath. You breathe in. Your body expands, and the water of the bath spills on to the floor. In the Earth's case, the oceans spill on to the continents, creating vast areas of shelf sea that can persist for millions of years.)

But even this phenomenon, which has been responsible for most of the great marine transgressions of geological history, is mere tinkering. If studies of fossil ocean floor from before one billion years ago are correct, at that time the ocean crust was over twice as thick as it was after that crucial moment in Earth history. This switch may have changed for ever the way the world looked and functioned.

How thick the ocean crust can be is governed by the volcanoes at mid-ocean ridges. The more active they are, the thicker the resulting crust (remember Iceland). On the other hand, the thickness of *continental* crust is a product of mountain-building processes, combined with the innate strength of continental rocks, which imposes an upper limit. When continents collide they build thick crust that extends both down into the mantle and up into the atmosphere. But there is a point above which mountains cannot rise, set by the limit of the rocks' mechanical strength. Beyond a certain weight and height, mountains are not mechanically strong enough to support themselves.

Bradley Hacker, Professor of Geological Sciences at the University of California, Santa Barbara, recently spent time investigating the Tibetan Plateau (which dates back 13.5 million years and is like a 'bow wave' to the collision of India with Asia). The Plateau affects weather worldwide. It plays a powerful role in creating the monsoons of India and Asia, for example, and has a global cooling effect on climate that may have helped tip the world into its current 'icehouse' regime shortly after it began to rise.

Although his was not the first contribution on this subject, what

Hacker confirmed was that although the crust thickens in the area of the collision, after a certain amount of thickening it weakens and spreads apart. He reported his findings in the journal *Nature* in 2001: 'Consider stacking pats of butter on top of one another. Imagine that stacking each pat . . . also generates heat, so that a thicker stack of butter is hotter than a thin stack.' In the case of the Earth, heat generated by radioactive decay within the rocks builds as the crust piles up, making the thickened crust weaker. Ultimately the rocks reach a certain maximum height and begin to flow outward. As Hacker concluded, 'There is a balance between the strength provided by the thickening of the crust and the weakness caused by heating from all that material.' The Tibetan Plateau is in a steady state. Currently standing at five kilometres tall, it will not get any higher.

For similar reasons of dynamic balance (and not including their very earliest days) the continents' average thickness has not changed very much through most of geological time. However, a pre-Rodinian world with much thicker *oceanic* crust would have been a very different place. All the ocean basins above the thick crust would have been much shallower. The difference between average continental height and average ocean floor depth, or 'continental freeboard' as it has been called, would have been much lower than now. Just as *The Book of Urantia* appears to have 'predicted', one billion years ago does indeed seem to have been an 'age of increased continental emergence'. Until the worldwide orogeny that created Rodinia, the amount of continental crust that poked above sea level would have been much smaller than it has been since. This is especially likely when we realize that, the mantle being very much hotter, more of the Earth's total water would have been in liquid form.

But as Rodinia formed, things were changing. The oceanic crust, responding to falling mantle temperatures, began to approach present-day thickness (about six to seven kilometres). Increased continental freeboard exposed more rock to the atmosphere, with a resultant increase in weathering on land. The chemical breakdown of rock materials sucked yet more carbon dioxide out of the atmosphere as it was converted to bicarbonate and carried away into the oceans in solution.

Because there was now more exposed land, seasonality also became more important on Earth than ever before, because land areas are much more susceptible to seasonal variation in the power of sunlight. Greater seasonality, combined with the return of more nutrients to the seas (as a result of enhanced weathering) further improved the oceans' organic productivity, which in turn led to even more carbon dioxide being swabbed from the atmosphere (just as even more algae could trap even more of it in even more lime mud).

The Earth's climate was reaching a threshold: a 'tipping point'. Rodinia's eventual break-up, on top of all these cooling factors, would precipitate the greatest climatic catastrophes ever to afflict our planet.

Within this cooling world Rodinia seems to have sat astride the Equator, leaving the planet's poles free of land, a rather rare event in Earth history. The stage was set. Rodinia gave way to the radiogenic heat building up beneath it, and started to fragment. Massive igneous provinces erupted, their dust and ash blocking out heat from the Sun, which by this point in its evolution was about 6 per cent weaker than today.

Supercontinents are arid because moisture cannot reach their interiors; but on smaller continental blocks this situation is reversed. After supercontinent fragmentation, more rain tends to fall on more land, and rock weathering speeds up. Because the continental fragments were then sitting entirely within the tropics, weathering rates were particularly high. What is more, the newly erupted basalt provinces were especially susceptible to chemical weathering.

So, as more rocks were weathered, even more carbon dioxide was removed from the atmosphere and delivered to the seas. The length of coastline increased, as did the area of shallow shelf sea, providing even more habitat for stromatolite-forming algae to colonize.

As these progressive effects took hold, they drove the climate into colder and colder territory. Icecaps began to form and expand. Normally, when icecaps expand over continents lying at high latitude, they exert a negative feedback on the process because they cover up more rock, preventing weathering and leaving more carbon dioxide in the air to keep the planet warm. But that didn't happen. With the continents far away from the poles, no brake was applied. As the icecaps crept Equatorwards to within about 25–30 degrees of latitude, they passed a point of no return. Earth was doomed to ten million years of icy stillness.

This point came because, at a certain coverage of brilliant ice and snow, the amount of heat reflected back into space became so high that the cooling process was unstoppable. The Earth system had no choice, the cooling effect had nowhere else to go but completion, encasing the whole surface of the planet in ice. This was how Lasagne World gave rise to Snowball Earth.

Iceworld

Once the planet was encased from pole to pole, the Earth system was frozen, literally and figuratively. Apart from rare nunataks of rock, the tallest mountaintops poking above the endless ice plain, the whole sunlit globe shone in only two colours: blue above and white below. On Iceworld there was no evaporation and no clouds anywhere. The hydrological cycle, in which water evaporates from the sea to be deposited as rain and snow on land and so returns in rivers to the sea once more, became restricted to the precipitation of the small amount of water that would 'sublime' from the ice surface: in other words, go straight from ice to vapour. And beneath their icy carapace, the seas became stagnant as interchange between them and the atmosphere ceased.

Yet deep below, far underneath ice, oceans and rocky crust, churning away irrespective of the catastrophe at surface, Earth's planetary heat engine rolled on, driven by the imperative to dissipate its radiogenic heat. Beneath the sealed seas, volcanically active spreading ridges pumped their acidic, superheated, mineral-rich waters into the frigid water; while above the ice, volcanoes that poked their hot heads through the frozen veneer spewed their gases into the atmosphere.

And there they stayed. No rain washed these gases out of the air and into the seas, and carbon dioxide began building up. It is a fairly safe uniformitarian assumption that Neoproterozoic volcanoes gave off at least as much gas as volcanoes today. From this it can be calculated that after a snowball event lasting ten million years, levels of carbon dioxide in the atmosphere would have risen possibly as much as a thousandfold. Earth's inner fire was about to save the world from the reign of the ice.

The end would have come suddenly. As the greenhouse effect kicked in, temperatures swung wildly upwards, to perhaps as much as 50°C at the ocean surface. Evaporation began again, further enhancing the greenhouse, since water vapour is one of the most powerful greenhouse gases of all. The water cycle now went into overdrive; torrential rainfall washed carbon dioxide out of the atmosphere, creating acid rain that landed on the newly exposed land surface (strewn with glacial rock flour) and so dissolved its minerals even more quickly. The reborn rivers returned huge quantities of bicarbonates to a sea already saturated by ten million years' worth of volcanic carbon dioxide pumped into it by submarine volcanoes.

Massive limestones were then deposited in shelf seas worldwide: seas that were also progressively deepening as the liberated water of the ice sheets filled up the ocean basins. Carbonates precipitated out on the seabed, depositing thick limestones directly above glacial deposits, with no sign of a time-break. Some of these limestones contain crystals of the calcium carbonate mineral aragonite, which normally only precipitates today from supersaturated pore-fluids as microscopic, needle-like crystals. Aragonite crystals that formed on the Neoproterozoic seabed took on gigantic dimensions as giant fans, some as tall as a man.

The ice-break of a snowball would have been Earth's most dramatic spring. The gradual release of the last Ice Age's grip 10,000 years ago must have been as nothing to the chaos that prevailed as the Neoproterozoic icecaps retreated. Persistent winds of over 70 kilometres per hour blew over much of the Earth's surface as a result of the vast air-pressure differences between the thawing tropics and the polar caps. These winds were not passing storms but lasted for years, causing huge oscillation ripples to form in sediments accumulating as deep as 400 metres: nowadays far beneath the reach of even the biggest storm waves. The ocean-surface waves generated by these global hurricanes probably exceeded seventeen metres in height, over huge tracts of sea. Below, meanwhile, the dissolved volcanic iron built up during the snowball finally oxidized and precipitated, and Banded Iron Formations suddenly made a comeback after a hiatus of a billion years.

Can all this really have happened? The geological evidence – glacial deposits at low latitudes, directly overlain by limestones, and the reappearance of BIFs – all fit the hypothesis. Moreover, they make sense of phenomena that have long been seen as 'anomalies', giving the Snowball Earth model immensely persuasive explanatory power.

As early proponents of continental drift had found, 'old-fashioned' geological evidence is qualitative, and often open to several possible interpretations. Powerful hypotheses, especially when they are 'non-uniformitarian' in the sense that they have no precise modern analogue on the planet today, are at once exhilarating and disturbing.

Doubters mutter that a powerful ruling theory is driving interpretation. Could the whole snowball story be no more than a 'selective
search after facts'? What of 'multiple working hypotheses'?

We have already seen some of the many different isotopes of
carbon, the element of life, and how their different atomic weights
lead them to behave differently in the natural environment. Carbon 12
is the common form, but there is an unusual heavy isotope, carbon 13,
which gains its extra unit of mass by having an extra neutron in its
nucleus. The carbon that comes out of volcanoes (mostly as carbon
dioxide) contains both isotopes in a well-known ratio. But any carbon
that has been involved in *life* processes has a different signature,
because photosynthesis, where everything begins, prefers carbon 12.
Living tissues therefore contain lower than average amounts of ^{13}C;
but conversely, limestones (principally calcium carbonate) that form
at times when life is thriving have above average values of ^{13}C because
they are made out of the carbon left over in the environment after life
processes have taken their share.

If you test the carbon-isotope ratios in limestones immediately
underlying and overlying the snowball's glacial deposits, remarkable
changes are revealed, bringing impressive support to the snowball
model. When pre-snowball limestones were laid down, life was thriving; so ^{13}C values begin high but drop steeply as the contact with the
glacial deposits approaches. This says that life was shrinking back
with the onset of snowball conditions, leaving more ^{13}C around in the
seawater to be incorporated in limestones. For ten million years or so
that the snowball lasted, no limestones were laid down. However, the
first limestones deposited after the snowball – the 'cap carbonates' –
remain low in ^{13}C because life had yet to recover. Then, gradually, ^{13}C
values rebound as resurgent life in the recovering shallow seas fractionated ever more ^{12}C into living tissues.

Between 710 million and 580 million years ago, the snowball cycle happened twice, possibly three and (some say) even four times during this never-to-be-repeated interval in our planet's history, as the supercontinent Rodinia split apart. So why did they stop happening? Why did snowballs not happen, for example, during our most recent glaciation, commonly known as 'the' Ice Age, which ended between ten and twelve thousand years ago?

The reason lies in the fact that, unique among supercontinents, Rodinia seems to have straddled the Equator, meaning that the world had no land at either pole. Never has that coincidence of low greenhouse gases, weak Sun and tropical concentration of landmasses come about again. Since the end of the Neoproterozoic, despite ice ages aplenty, none has ever gone to anything approaching a snowball. Since the break-up of Rodinia, the safety catch has been back on. Iceworld was finished.

Snowball or slushball?

In structuring the story as I have, however, I have taken one particular route through a mass of scientific evidence treading a line of stepping stones across a torrent of argument. The Snowball Earth hypothesis remains controversial and contested.

Take, for example, the one big question mark over the whole idea of the snowball model as advocated by Paul Hoffman and his supporters. How could a total white-out, involving a global ice-cover perhaps many kilometres thick, ever have allowed photosynthetic organisms in the oceans to survive? Clearly, life *did* survive successive snowballs and therefore, argue the theory's critics, each so-called snowball must really have been a 'slushball', preserving ice-free refuges at the Equator.

The slushball model, on the other hand, produces a slower deglaciation, with no sudden 'flip' from icehouse to greenhouse, and has atmosphere and hydrosphere remaining in balance throughout. A Slushball Earth might not even have had uniformly anoxic oceans. Can this explain the recurrence of BIFs? Some geologists have found evidence for erosion and deposition, with glacier advance and recession at this time. Those who believe in the Hard Snowball have to say this all took place during the short deglaciation phase, because their model predicts a total shutdown of such hydrologically dependent activity for ten million years. This may be true; but is it?

Could computer models help resolve these issues? They tend to produce a runaway snowball effect when levels of atmospheric carbon dioxide are about the same (or only slightly lower) than today. However, not all climate computer models display this kind of flip-flop instability. For example, models in which no heat is transported across the ice-line are less likely to go to total snowballs because the tropics stay warm enough to remain melted. Programs that couple atmosphere and ocean circulations are also resistant to global glaciations, because they allow ocean convection at the ice-free tropics. So climate models, on their own, do not provide conclusive evidence because they can be tweaked, within quite reasonable bounds, to support a number of plausible outcomes.

So although models that result in 'slushball' solutions keep biologists happy, they do not appear to explain the geological evidence, especially the 'cap carbonates', and the temporary return of BIFs, quite as well. The full snowball model also demands that deglaciation be very rapid, which is consistent with the way cap carbonates seem to have been deposited, without any time gap, on top of glacial deposits.

Resolution of this impasse hinges on one crucial question: exactly how thick did the ice of Snowball Earth get? We know how sea-ice thickness and surface temperature are related in modern oceans, and we know from computer models roughly how cold it would have been

during a snowball episode. Applying these simple formulae in a uniformitarian way suggests that, during a full snowball, ice at the tropics
should have exceeded a kilometre in thickness, far too thick for any
light to get through. How, under a full snowball, could slimeworld
have survived even *one* ten-million-year-long night, let alone two (or
more)? Clearly, somehow it did.

In 2000 a new suggestion came along to break the impasse, involving a more biology-friendly 'thin ice' model. In this model the
ice-cover is total, but just a few metres thick at the tropics: thin
enough to allow light through while providing enough of a seal to
restrict the hydrological cycle to a minimum and prevent the oceans
breathing. The idea came from David Pollard and James Kasting, of
the Earth and Environmental Systems Institute at Pennsylvania State
University.

With ice, thinness and transparency go hand in hand. Opaque ice
is compelled to be thick. You could call this another 'greenhouse
effect', because transparent ice, on the other hand, traps heat like a
greenhouse's glass panes, melts itself from below, and stays thin.

Today's sea ice is full of inclusions that scatter the light and make
it highly opaque. Critics of the 'thin ice' idea were quick to point this
out; for if ice at the Neoproterozoic tropics was like modern sea ice,
it would have been opaque, hence also thick. However, Pollard and
Kasting are not so sure about this uniformitarian approach.
According to them, the ice at the tropics would have formed by a
combination of two processes: the Equatorward flow of ice from high
latitudes, forming 'sea glaciers', and water that simply freezes on to
the bottom of the sheet.

When sea ice grows in the modern-day Baltic, for example, water
freezes to the underside of the ice sheet, trapping pockets of brine
that make the ice opaque. But Pollard and Kasting's model suggests
that Neopoterozoic Snowball Earth ice would have formed at a mere

seven millimetres per year – much more slowly than modern sea ice. Such slow freezing would have produced much clearer ice.

As for 'sea-glacier' ice, its nearest analogue today is the ice seen on land glaciers, like those in Antarctica, which form by the accumulation of snow. In land-glacier ice the main light-scattering inclusions are bubbles of air, which originally lay between the snowflakes before they were annealed together by pressure. According to Pollard and Kasting, for Neoproterozoic sea-glacier ice to have been clear enough for it to stay thin, it must have had a bubble density of no more than about 0.32 per square millimetre, and that lies well within the range of bubble densities seen in the upper parts of ice cores taken from Antarctic ice sheets today.

So it appears that life could indeed have survived ten million years in the chiller, because when the freezer door closed the light didn't, for once, go out. Slushballers, on the other hand, regard the thin ice idea as an unnecessary sophistication. They do not believe that total ice cover is required to ensure that the oceans stagnate and so accumulate ferrous iron in solution. Evidence emerged in 2005, from organic-rich rocks dated to 700 million years ago, that suggested to Alison Olcott of the University of Southern California in Los Angeles and her colleagues that not only was photosynthesis operating during the snowball, but it was widespread, tropical and happening in stagnant water. If their interpretation of the biomarkers in these rocks from south central Brazil is correct, even the photic zone of the Neoproterozoic ocean was oxygen-free. Perhaps there *was* a thin ice cover, as Pollard and Kasting suggest; but perhaps there was simply no ice at all. Perhaps a tropical, ice-free waistband around the Earth was just not enough to break up the stratification of the global ocean.

As I finish writing this book at the beginning of 2006, another international conference, this time in Switzerland, is planned for the summer. More evidence, from all over the world, will be presented in support of new interpretations of these pages from the greatest palimpsest that may settle the controversy between snowball and slushball. But just as Lemuria finally sank beneath the waves of new knowledge, today's closest approximation to truth slides into myth as the latest ideas are subjected to the evolutionary pressure exerted by the realities of new evidence.

Hope and glory

Can it be entirely coincidental that, after three billion years during which the pinnacle of evolution was green slime, complex life burst into existence just as the last snowball melted away never to return? Could it be that, if a low-latitude supercontinent called Rodinia allowed the snowballs to happen, and if the snowballs somehow gave life the kick in the genes it needed to develop complexity, Rodinia really *was* our motherland? Could the vagaries of the Supercontinent Cycle be the main reason why, in place of universal lasagne, we have in Jacques Prévert's words, 'New York passions, Parisian mysteries, the little canal at Ourq, the Great Wall of China, the river at Morlaix, legionnaires, torturers, rulers, bosses, priests, traitors, pretty girls and dirty old men'? Could Rodinia be the reason we have such a diverse world? Could Rodinia be the ultimate reason we are all here today, doing what only humans do: wondering how we got here?

To try to answer this question, we need to know something rather accurately. When exactly did complex life first develop? Only when we know that can we hope to judge whether there is a case here to answer.

On 12 November 1931, three years before his death at the age of seventy-six, Sir Edward Elgar and the London Symphony Orchestra performed the trio from the *Pomp and Circumstance March No. 1* ('Land of Hope and Glory') for the opening of EMI's new recording studios at 3 Abbey Road, not far from a certain pedestrian crossing later made world famous by the Beatles.

The Abbey Road recording was not Elgar's first foray into this newfangled technology; he made his first recording in 1914, weeks before the First World War broke out. But despite the poor quality of contemporary reproduction (imagine a wind-up gramophone playing 78rpm shellac discs using a needle and a big horn as acoustic amplifier), these ancient recordings actually contain a wealth of sound detail that was invisible – or rather, inaudible – to the old technology. With digital remastering all kinds of unimagined detail can now be heard. All that information was always there, but only the new tools allow it to be revealed. The same is true of the geological record.

When geologists began looking systematically for fossils for the first time in the nineteenth century, using William Smith's discovery that you could identify and correlate strata of any given age by the fossils they contained, they noticed that rocks below the Cambrian system were barren. The term they gave to the whole 'Cambrian and younger' geological record was 'Phanerozoic', which means 'evident life'. The apparent suddenness of life's appearance posed a great problem for evolutionary theory because at 542 (plus or minus one) million years ago, and seemingly from nowhere, nearly all the main animal body plans (arthropods, molluscs, echinoderms and so on) seemed to burst on to the scene and hit the ground running, swimming and burrowing like there had been no yesterday. It all gave Darwin sleepless nights.

Since then, geologists' reading of the record has become more sophisticated. Now, instead of just looking for things they can hold in their hands, they can detect fossils and fragments of fossils mere microns across. Using the tools of organic chemistry, they can even pick up the molecules of life, so-called 'biomarkers'. These chemicals, many of which are quite specific to certain types of organism, are remarkably durable in the fossil record. Now, when apparently barren rocks are ground up, dissolved in organic solvents and passed into a mass spectrometer or a gas chromatograph, these telltale molecules stand out as diagnostic peaks in the instrument's read-out: the merest whiffs of long-vanished life.

Darwin often invoked the imperfection of the fossil record to get himself out of such difficulties with evidence, and he was quite right to do so. The circumstances under which a fossil can form are rare. The nineteenth century's fossils, the hand-specimen or 'macrofossil', is a very scarce beast. Microfossils and molecules, on the other hand, have a much higher chance of being preserved, and so provide a much more reliable tool for judging 'first appearance'.

Think about it: if an organism evolves at a certain moment, it will take some time for this creature to become common. Yet even its hard parts (its shell, or bones) stand a very slim chance of being preserved as fossils at any time, let alone during that species' very earliest days on Earth. Pile upon these slim chances the chance of those rare fossils surviving the vicissitudes of all subsequent geological history and then add the further unlikelihood of their being *found*, and you produce some very unfavourable odds indeed. So, any macroscopic species' first appearance in the fossil record is bound to post-date its true first appearance on Earth.

But microfossils are different. Microscopic things are everywhere in the environment – ask a hay-fever sufferer. We are trying to produce an accurate date for the first appearance of complex multicellular animals, when the first actual fossils we may discover will post-date that event quite considerably. It would be very useful if the creature in question produced something durable and microscopic in astronomical numbers. Unfortunately, unlike modern plants with their spores or pollen, animals don't do this.

Alternatively, you could test for animals' first appearance by using some superabundant microfossil as a proxy. It is a reasonable assumption that the appearance of multicelled animals had profound ecological effects, and that these might be visible in the remains of other organisms. It was, after all, the first time any of them had ever been eaten. You would expect this to provoke an evolutionary response. You might therefore be able to detect something both in the appearance of microfossils (a change to the roughness and durability of their armour-plating, for instance) and in the style and pace of their evolution.

Such a potential proxy group exists in the rather unprepossessing form of tiny organic sacs called acritarchs. Acritarchs are an ancient but artificial grouping of microscopic (20–150 microns across), organic-walled fossils found in nearly all sediments – once you have dissolved away everything else in bath after bath of strong acids. Acritarchs are thought to represent the 'resting cysts' of single-celled algae with a many-staged life cycle. They were first discovered in 1862, but the term 'acritarch', which just means 'of uncertain origin', was only coined in 1963. Acritarchs are useful for correlating sedimentary rocks of Proterozoic and Palaeozoic age mainly because they were the only microfossils around then; but their usefulness as correlation tools increases enormously after the snowballs.

The oddest thing about acritarchs is that before the younger of the two main snowball events, the Marinoan glaciation, individual acritarch species tended to exist for 1000 million years: something inconceivable in the modern biosphere. But after this period of extreme evolutionary stasis, at about 650 million years ago, everything changes. Thereafter, acritarch species tend to survive only for a few tens of millions of years. Something fundamental in the way the biosphere worked had changed.

This is thought to be the proxy evidence pinpointing the moment that planktonic animals appeared. For the first time, the acritarch-producing algae found themselves to be prey. Before then, the seas had probably been a saturated algal soup, with nothing more than the availability of nutrients to control runaway growth. No wonder things did not change for a billion years. Why should they?

But now things were different and soup was on the menu. Life had turned on itself for the first time, and the selection pressure this imposed accelerated the origin and the extinction of new acritarch species. The spiny, heavily armoured and relatively short-lived forms of the post-snowball world are nothing less than the first tooth and claw marks of nature's new order.

A little later, at around 580 million years ago, we find the first unequivocal animal fossils, in the form of a beautifully preserved clusters of cells representing embryos of an animal (though we cannot say what sort), in the mid to upper parts of the Doushantuo Formation of southern China. Not long afterwards the first trace fossils – those telltale marks made by animals moving across and through the sediment – first appear, in rocks no younger than 558 million years old, in the Verkhovka Formation of north-west Russia.

The period we are speaking of is called the Ediacaran, now officially named for the remarkable forms over which Mark McMenamin and many others have been puzzling. The Ediacarans were part of this new evolutionary order, but although some may have evolved into animal forms that are still around us today, others (perhaps, who knows, those photosynthesizing animals in McMenamin's vision of the Ediacara Garden), after first trying sheer size as an evolutionary refuge against the mounting pressures against them, finally succumbed. They succumbed to the destruction of the algal lasagne they rested on and ended up as lunch for those voracious new predators that, tiring of soup and lasagne, moved on to the meat course.

Code breakers

There is one final line of evidence for life's sudden dash to complexity after the snowballs, suggesting that these climatic crises were indeed what set the process off. That evidence is found inside the molecule of life itself. DNA, the molecule that carries the genetic codes and governs our growth and development by regulating the process whereby amino acids build proteins, is vast; yet it is a lot vaster than it needs to be, since large tracts of every organism's genome have no known function. Because they are not expressed in the biology of the organism, they are not subject to natural selection like other areas of the molecule.

Natural selection is often thought of as what 'changes' organisms, but this is only true if something in the environment is changing, disturbing the equilibrium. Under stable conditions, natural selection is a strongly conservative force that keeps things the way they are and makes sure that things that aren't broke *don't* get fixed. But those areas of the DNA molecule that are not expressed biologically are able to mutate freely without impairing an organism's evolutionary 'fitness'. Biologists have found that, left to its own devices in this way, this so-called 'junk' DNA mutates at a rate that (over geologically long periods of time) is constant enough to be used as a clock (though it has to be regulated by reference to a good fossil record).

Despite what the film *Jurassic Park* would have you believe, very little DNA is preserved in the older fossil record, and the further back you go, the less survives. So if you want to use DNA to date events that took place more than 500 million years ago, finding DNA from those times is not an option. But fortunately there's no need; we can look at the DNA of different groups of *living* animals (sponges, worms, molluscs etc.) instead. Then we can combine what we know about the rate at which the DNA clock ticks, with a statistical expression of the difference between the junk DNA sequences of sponges,

worms or molluscs, to work out how far back we would have to
extrapolate these in order to make the different groups' DNA look the
same. The result should be the point in time at which the groups' evolutionary paths diverged, and it should be confirmed by the fossil
record. On the supercontinent of science, everything must fit together.
The molecular biology of today's animals confirms the fact that they
are all related, that some groups split before others and that every
living thing is the product of its unique history: a history that dates
back billions of years and ultimately makes us one with the stars.

To confirm whether the snowballs and the Supercontinent Cycle
might have been responsible for creating complex life, the crucial evolutionary divergence that we want to date is the one that gave animals
the power to move purposefully, even *through* sediment (burrowing),
to ingest material, process it in a gut and expel excrement. For these
were the developments that changed nature for ever and finally dug
up the Garden of Ediacara. This evolutionary event was all to do with
how the tissues of animals became organized and differentiated in the
embryo.

Sponges are simple multicellular animals, but they are little more
than balls of barely differentiated cells. Jellyfish, on the other hand,
are two-layered animals. This allows them to move, albeit in a rather
undirected way, as they drift with currents or to move away from
harm (they hope) when warned by their rudimentary sense organs.
They have an outer layer of cells and an inner one, separated by a
springy, gelatinous mass. Muscles around a jellyfish bell contract
against the springiness of this mass, which bounces back into shape
when they relax, allowing the swimming cycle to start again.

But the next development in the history of embryology was crucial
because it created animals with *three* layers of tissues, the middle layer
developing from the jelly-like stuff as it became invaded with cells.
Moreover, the embryos of the new three-layered animals developed a

method of 'turning in' on themselves. First, the embryonic ball of cells became hollow, to create the body cavity. A fold in the outer layer then developed, forming a pouch on the inside. This became budded off along most of its length, but remained connected to the exterior by a single pore that developed into, literally, the fundamental opening: the anus. The mouth developed later, at the opposite end. You can see this evolutionary process re-enacted every time a human embryo develops today, and it is the reason why you have a body cavity filled with tubular guts.

Three-layered animals were a giant leap. The outer layer gave skin, nerves, ears and eyes. The middle layer gave rise to muscle, bone and the circulatory system; while the inner layer created the digestive, glandular and respiratory systems. Worms, snails and humans all have this same basic organization. Having three layers meant having more cells per unit volume, and separating the digestive tube from the body wall created problems of transport. Oxygen needed to be brought in, and wastes and nutrient carried out. All this required ducts, circulation systems and respiratory mechanisms that simpler animals didn't need because they could live by absorption alone.

Above all, three-layered animals were directed by sense organs concentrated at a head end, and their muscles enabled them to move in more or less any direction in active pursuit of food. The mouth, being also at the head end, was the first thing to arrive at the food source, ingesting what it found, filling the gut with a mixture of food particles and sediment. Burrowing and grazing modes of life were born; and as a result, trace fossils started to appear in the fossil record, and the fine, undisturbed laminations of Lasagne World became rare as the Garden of Ediacara went under the plough.

According to the molecular clock, the first three-layered animals, complete with heads and nerves and muscles and guts, should have appeared about 580 million years ago, which places this evolutionary

event right in the middle Ediacaran (630–542 million years); just after
the Marinoan snowball had melted, after its cap carbonate had been
safely deposited, and the climate had settled down again. What is
more, 580 million years also turns out to be the date when fossils of
the Ediacaran animals, though already doomed, became abundant
enough to be preserved in non-exceptional circumstances. And all this
took place barely forty million years after the last of the two major
Neoproterozoic white-outs came to an end.

As scientists are fond of pointing out, correlation need not imply
causation; but to many, these convergences are too consistent and too
numerous to be meaningless coincidence. If true, the correlation
between the origin of complex life and the end of major snowball
episodes firmly ties our own origins to the Supercontinent Cycle,
because it was the chance siting of continents around the Equator
1000 million years ago that made those all-important snowball events
possible.

Life itself probably owes its origin to the geology of ocean-floor
hydrothermal vents. But that part of it that isn't slime may owe its
brief 580-million-year tenure of this planet to nothing more than
random turbulence within the Earth's convecting mantle, which once
swept the continents to the tropics. There, far from stopping a runaway snowball, their fragmentation from maximum packing
enhanced weathering, created volcanic cooling and multiplied the
length of shallow, lasagne-covered sea floor, just at the time when a
billion and a half years of photosynthesis and carbon sequestration
had already made the Earth System especially vulnerable.

Out of this catastrophe came glorious, complex life. And now, we
are blessed (or cursed) with the brains to work it all out for ourselves.
The question then becomes: will we bother, or will we sit, like
Albrecht Dürer's *Melancholy*, surrounded by the tools that can help
us explain the world, but too indolent to use them?

Melancholia by Albrecht Dürer. Hapless Melancholy lies surrounded by the untouched tools of science and art.

EPILOGUE

LIFE, THE UNIVERSE AND THE PUDDLE

For nature, heartless, witless nature,
Will neither care nor know

A. E. HOUSMAN, LAST POEMS XL

On the morning of Sunday, 26 December 2004 about 1300 tourists and pilgrims crowded on to Kanyakumari rock, a charnockite islet 200 metres or so off the southernmost tip of India. Geologists call the islet 'Gondwana junction' because it marks the 550-million-year-old suture where India, Madagascar, Sri Lanka, East Antarctica and Australia once joined together to build the eastern portion of Suess's Gondwanaland. But to followers of Vedantist spiritual philosopher Swamy Vivekananda (1863–1902), the rock is remarkable for being the foundation for his memorial, which opened to visitors in 1970. This was the impressive structure that drew them to Kanyakumari as the sun came up over the Bay of Manar that morning.

Mr G. Ramalingam of the Port Authority remembered afterwards that they had sold 3469 tickets for sailings between 8am and 9.45am, though they had suspended sailings to another outlying rock, Valluvar, at 9am – a decision that probably saved about 500 lives because the monument to the great Tamil poet, a forty-metre-high statue on a square plinth, offers no shelter. For 26 December would prove to be a much more memorable day than anyone expected.

Although not one of the tourists and pilgrims would die at Kanyakumari, some 230,000 others around the Indian Ocean would lose their lives before the day was out.

Perhaps as they were preparing for their expedition, some of those pilgrims may have been aware of a slight earth tremor; but most of them either didn't notice or slept through it. Yet even as that distant seismic shock rumbled through India and around the world, slower and much more deadly waves began spreading across the Indian Ocean. By the time the pilgrims climbed aboard the ferry of the Poompuhar Shipping Corporation that would take them to the island, tens of thousands were already dead in Indonesia. Thousands more lives were being lost in Sri Lanka, just over the horizon. Soon the wave would turn the corner and sweep up Sri Lanka's west coast and bear down upon Kanyakumari.

No surprise

The deep ocean trench that skirts the Indonesian Archipelago on the other side of the ocean, marks the contact between two of the tectonic plates making up the cracked eggshell of the Earth's crust. One is the Australian Plate, consisting of Australia and the floor of the Indian Ocean, and the other, to the north, carries Europe and Asia and is called the Eurasian Plate. At this trench the floor of the Indian Ocean is subducting, sinking down into the mantle, beneath the island arc of Indonesia. This is but one small part of the long process of building the next supercontinent, piece by piece, each fragment edging into place, just as India has already been annealed to Asia in the collision that is today creating the Himalayas and the Tibetan Plateau.

In many ways the earthquake that caused the 26 December tsunami should have taken nobody by surprise. There are known to have been two great earthquakes of over magnitude 8 along this part of the Indonesian Arc: in 1833 and 1861. The zones of rupture that caused these two events sit along adjoining, non-overlapping parts of the same plate boundary, adjoining the Batu Islands. No quake of similar size happened during the twentieth century, until June 2000, when a 7.9 quake struck near Enggano at the extreme south-eastern end of the 1833 rupture zone. The 26 December event extended movement along the plate boundary from the island of Simeulue, at the other end of the chain, almost to the coast of Myanmar (Burma). This left a gap of a few hundred kilometres between Banyak and Simeulue over which no movement at all had taken place in historical time. The omission was rectified on 28 March 2005, when the last 'stuck' part of the fault gave way in an 8.6 magnitude tremor that thankfully produced no very serious widespread tsunami.

The processes of plate movement are not smooth at our human timescale; the heat engine of the Earth is no perfectly oiled machine. So while the average rate of convergence between South-East Asia and the floor of the Indian Ocean may be 5.2 centimetres per year, unlike your growing toenails, these figures represent the averages of many sudden discrete movements, some of which, especially when long delayed, can be very large and very sudden indeed.

On 26 December 2004 stress that had built up over hundreds, perhaps thousands, of years was finally released along a 1200-kilometre stretch of plate boundary to the north-west of Simeulue. This stupendous quantity of energy, stored as though in a giant leaf-spring, was liberated in seconds as the crust of the overriding plate (which had been bent downwards by the pressure of the sinking ocean floor beneath) bounced back, flinging the ocean up by as much as ten metres.

At a magnitude of 9.3 (on a scale whose every increment denotes a quake thirty times larger than the one below), this was the largest

The Sumatra subduction system, showing the dates of known historical earthquakes and the sections that moved during each. First published in *Geoscientist* 15, 8 p. 4. Reproduced courtesy of Dr John Milsom.

quake unleashed by the Earth for half a century, the second strongest ever recorded, and the first true 'Global Geophysical Event' since Krakatoa erupted in 1883. Its seismic waves, travelling quickly through the Earth's crust, crossed the Indian Ocean and passed through the Vivekananda Memorial almost instantaneously, as the whole planet reverberated like a bell struck with a massive hammer.

The tsunami meanwhile rolled outwards from where the ocean floor had been uplifted. In deep water, travelling at the speed of a jumbo jet, its waves were only a few centimetres high, passing unnoticed beneath the hulls of container ships crossing the Bay of Bengal, stacked and sleeping in the morning light. But in the shallows, the waves slowed and bunched together, piling up huge cliffs of surging, turbid water that rushed inland like supercharged high tides, sometimes tens of metres tall, razing all before them, scouring the coastline of an entire ocean.

Rescue

The tourists and pilgrims, now disembarked at the Vivekananda Memorial, watched as the horror unfolded. The morning was calm and the sky clear and blue. The first thing the visitors noticed was the withdrawing roar of a false tide, as though someone had pulled a plug on the ocean. Dark, wet rocks at the foot of the many islets, and finally the seabed in between, were suddenly exposed. It was as though the sea had inhaled. In the eerie quiet, which had almost silenced the chatter on the Memorial, the visitors could hear the hiss of air being sucked into the pore spaces of the draining sand and the flapping of a few stranded fish. Then, just as the onlookers had begun to shrug their shoulders at the sight, a series of huge waves, each several metres high, rushed in, crashing over the sunstruck promenades surrounding the Vivekananda Memorial. The great statue of Tiru Valluvar was engulfed in spray, like a deep-sea light breasting an Atlantic storm, but on a cloudless morning.

How many of the visitors to that tiny outcrop of charnockite thought at that moment about Katalakōl, of the lost books, and the palaces of the great scholar kings who, in the dreamtime of Tamil myth, held benevolent sway over the lost lands of Ilemuriakkantam? Perhaps many thanked the gods, because they had indeed made a life-saving choice that morning. All three boats of the Poompuhar Shipping Corporation were washed ashore by the tsunami; but after a long wait all the visitors were rescued, not by the Indian Air Force helicopter, which found it could not land, but by local fishermen whose boats survived the tsunami and who made several sorties to pluck the visitors to safety.

Over the days and weeks that followed, many stories emerged about how, here and there around the ocean, a little learning had been a life-saving thing; tales of the teacher who saw the tide go out unexpectedly and shooed all her pupils upstairs just in time. They served to show the life-saving power of knowledge; knowledge that most simply didn't have. But it became clear that, given the right combination of technology and education, many might have enjoyed a fate like that of the lucky fishermen of Nallavadu, on India's eastern coast.

The story goes that the son of one Nallavadu fishing family was on holiday in Singapore when he saw a news report of a massive earthquake and rumours of a terrible wave. He telephoned his sister and told her to spread the word and leave home immediately for high ground. This small community had for several years benefited from the presence of a small Internet-linked communications centre, set up by the M. S. Swaminathan Research Centre in Chennai to provide information to fishermen about weather in the Bay of Bengal. Armed with the news, villagers broke into the centre's telecoms facility and, using its public-address system, told the village's 500 families to run for their lives. And in the end, not one life was lost from Nallavadu's population of 3500; though 150 houses and 200 boats were reduced to rubble and matchwood.

For scientists, especially those working at the Pacific Tsunami Warning Centre in Hawaii, the unfolding situation was immensely frustrating. Together with seismologists all over the world, they had detected the massive quake and knew that a tsunami was a likely consequence. But there is no regular correlation between earthquake magnitude and tsunamis; and without any tsunami sensors in the Indian Ocean, still less any established lines of communication with the countries bordering it, it was impossible for them to get any warning to those who might have benefited (except, finally, to the Horn of Africa, where casualties were low as a result).

But it was not long before the political will arising from the disaster began to take effect. On Monday 10 October 2005 a German research vessel set sail from Jakarta to place the first of fifteen earthquake sensors on the seabed some 620 miles offshore. Attached by ties to large buoys at surface, their signals are now being continuously beamed to the offices of Indonesian government geologists, and warnings can be relayed to the media and the public via SMS, fax and email. For its part, India decided to set up a tsunami warning centre in Hyderabad at an estimated cost of $27 million, and by December 2005 an interim Indian Ocean tsunami early-warning system costing $53 million, tying together seabed earthquake sensors and tide gauges, was nearing completion thanks to the United Nations Intergovernmental Oceanographic Commission (IOC).

As usual, because such warning systems involve links at all levels stretching from international to local, coordination will be the biggest problem. Emergency preparedness planning, awareness campaigns, drills and local evacuation plans, educational programmes, and installing emergency operational capability – all these need to be present across a vast area that pays no heed to national boundaries. But without these things, none of the new technology will prove to be any use at all.

Useful knowledge

Science historian Naomi Oreskes has written: 'Scientists are interested in truth. They want to know how the world really is, and they want to use that knowledge to do things in the world.' It was this same impulse that drove Eduard Suess to design and build his clean-water scheme for Vienna, or Henno Martin and Hermann Korn to find water for Namibia, or John Joly to apply radiotherapy to the treatment of cancer.

Earth scientists often complain, with reason, that politicians underuse the full potential of their subject, especially for the benefit of vulnerable (for which read 'poor') people living in unsafe housing in unstable places. But, in times like the tsunami's aftermath, this feeling rises to a pitch higher than mere frustration. That feeling is *despair*: that the world is still so ruled by the short-term, by superstition, inertia and irrationality, and that their humane, possibilist long-term view of the world is not only ignored but even denied.

If today there is fresh water on Namibian farms and in Vienna, and an emerging tsunami early-warning system in the Indian Ocean, it is because geologists in the past have done the science that brings a closer understanding of deep time and the inner workings of the Earth. You cannot pick and choose with science. A seemingly rarefied geology that reconstructs the lost supercontinents of Earth's deep past is the same science that (with political will) can save hundreds of thousands of lives in the Indian Ocean when the next tsunami strikes. The arcane business of how our Earth's atmosphere evolved during the Precambrian under the influence of evolving life is the same science that now helps us understand the massive, uncontrolled climate experiment in which the human race is currently engaged. But to deny one part of science is to deny it all. Science hangs together. It is a supercontinent.

It is also progressive, as its ideas approach ever more closely the actual truth of nature as revealed in the great palimpsest of the geological record. 'Progress' may be an unfashionable Enlightenment notion, but in science it is real; and the test of that progress's reality is the ever-increasing power that science puts in our hands. Just as the history of the Earth is made up of both repetitive cycles and directional arrows, as the wheels of science turn, throwing up the same ideas time and again throughout intellectual history, the train to which they are fixed moves forward.

When thy judgments are in the Earth

Therefore how grotesque was it to read, just seven days after the tsunami struck, in the *Sunday Telegraph*, whose front pages were given over to detailed geological explanations of the earthquake and tsunami, of a new folly being made ready for its first visitors in Petersburg, Kentucky, USA. Called the Museum of Creation and costing about the same as Hyderabad's tsunami early-warning centre, the theme of this particular park is the literal truth of the Old Testament creation myth, which it seeks to uphold against all (genuine) scientific evidence. Just as the Tamil devotees appeal to outmoded nineteenth-century science to bolster the idea that their national myth is literally true, here the Old Testament creation story is bolstered by what the museum's backers call 'creation science'.

This non-subject, devised by young-Earth creationists to lend credibility to their prejudices, is alas much more than some regrettable but harmless local dispute about the romantic tales of ancient poets. Overenthusiastic appeals by Tamil politicians to a few outdated science references may occasionally be embarrassing for their academics; but it remains, at most, a little local difficulty. On the other hand, the purpose of 'creation science' is to misrepresent real knowledge in a crusade to replace free enquiry with slavish adherence to simplistic dogma – with belief in the Word before the world.

I have tried in this book to show something of how ideas in science often grade into – perhaps even sometimes derive from – ideas in myth, and I have done this to show how important it is to know the difference between the two. The truth is that we, as a species, can no longer afford the luxury of irrationality and prejudice. We are too many and too powerful to live in dreams. And the greatest and most irrational of the prejudices from which we must free ourselves is one identified by Lucretius in the last century BC: the belief that the world was made for us.

The supercontinent story tells us, like no other in Earth science, that she was *not* made for us – any more than she was made for the trilobites that grubbed around in vanished Iapetus, or for the *Glossopteris* tree or the little *Mesosaurus,* whose fossils reunited Gondwanaland, or the tiny feeding-trace *Oldhamia*, on whom John Joly mused. Douglas Adams picked up this theme in what I call his 'parable of the puddle':

> . . . imagine a puddle waking up one morning and thinking, 'This is an interesting hole I find myself in. It fits me staggeringly well; must have been made to have me in it!' This is such a powerful idea

that as the sun rises in the sky . . . and the puddle gets smaller and smaller, it's still frantically hanging onto the notion that everything is going to be all right because this world was built to have him in it; so the moment he disappears catches him rather by surprise. I think this may be something we need to be on the watch-out for . . .

We can, if we choose, either fret over our lost futurity or comfort ourselves with the thought that one day our species may shuck its bonds and spread throughout the galaxy; and that our space-going descendants may, millions of years in the future, rediscover our home planet after the greatest racial diaspora of all. Maybe that way, our direct offspring will see the next supercontinent on Earth. But this is a long shot. Until we can live without her, Earth is not a part of *our* story – *we* are a part of *hers*.

As the poet Hugh MacDiarmid put it:

> *What happens to us*
> *Is irrelevant to the world's geology*
> *But what happens to the world's geology*
> *Is not irrelevant to us.*

The last dethronement

Science has been trying to humble the hubris of humans from the start, in a series of what Sigmund Freud referred to as 'dethrone-ments'. The first dethronement was of the Earth as the centre of the universe. Second was our own dethronement as a unique creation in the image of God. Third (in Freud's opinion) was his demystification of the human mind's deepest motivations.

Science is not often thanked for delivering such slights to our collective ego; though in fact these blows have been nothing like crushing enough. For when, like Douglas Adams's puddle, we find ourselves standing on the brink of destruction it will be our arrogance, as much as the ignorance on which it feeds, that will prove our undoing.

Science cannot tell us everything that matters about being human, but it provides us with the only practical knowledge of the natural world in which we have any reason to believe. We know this because it works. But science also teaches us another important lesson – that there is no absolute knowledge of any kind – either about the Earth, or anything else. True, science can put some things past reasonable doubt: organic evolution or the age of the Earth are now well beyond that point. Despite what they may tell you in the Museum of Creation, the likelihood such basic scientific ideas being simply wrong is precisely nil. But the key word here is *reasonable*. Nothing ever remains beyond unreasonable doubt, especially to the fanatical adherents of outworn creeds who desire only to enslave.

The discovery of deep time is perhaps the greatest single liberating contribution that Earth science has made to wider culture. Conceiving of a timeframe large enough to encompass many repetitions of a cycle that can span 500 million years or more changes one's perspectives – especially on how properly to judge the relationship between ourselves and the Earth. As our species becomes more numerous and powerful, our last chance of long-term survival will depend on embracing yet another dethronement. We have to realize that we are the puddle, at the mercy of circumstances, but at least able to figure out how to keep ourselves alive and comfortable if we use the capacities with which evolution has equipped us.

Lucretius, speculating about the age of the Earth, came to the mistaken conclusion that it was new. For if not, he asked, where were the works of the poets who sang before Homer? Twenty centuries later John Joly wrote in reply:

We do not ask if other Iliads have perished; or if poets before Homer have vainly sung, becoming a prey to all-consuming time. We move in a greater history, the landmarks of which are not the birth and death of kings and poets, but of species, genera, orders. And we set out these organic events not according to the passing generations of man, but over scores or hundreds of millions of years. We are . . . in possession today of some of those lost Iliads and Odysseys for which Lucretius looked in vain.

FURTHER READING

Books

Adams, Douglas, 2002. *The Salmon of Doubt*. Pan. 284pp. Posthumous collection of writings by the renowned author of *The Hitchhiker's Guide to the Galaxy*. Contains the 'parable of the puddle' in the essay 'Is there an artificial God?'

Benton, Michael J., 2003. *When Life Nearly Died – The Greatest Mass Extinction of All Time*. Thames & Hudson. 336pp. Readable textbook by an acknowledged expert and popularizer, focusing on the end-Permian extinction and possible reasons for it.

Bronowski, J., 1973. *The Ascent of Man*. BBC. 448pp. Accessible and authoritative examination of the place of science within the rise of human civilization.

DeCamp, L. Sprague, 1954 (rev. 1970). *Lost Continents: The Atlantis Theme in History, Science and Literature*. Gnome Press (1954), Dover Publications (1970). 348pp. Spirited account of the influence of Plato's Atlantis story on subsequent lost-world makers.

Greene, Mott T., 1982. *Geology in the Nineteenth Century: Changing Views of a Changing World*. Cornell University Press. 324pp. A classic, well-written analysis of Earth science's heroic age, including the origins of drift theory.

Koertge, Noretta (ed.), 1998. *A House Built on Sand: Exposing Postmodernist Myths About Science*. Oxford University Press. 322pp. A selection of essays by scientists and philosophers that examine postmodern constructions of science and analyse the political damage they inflict upon science in society.

McMenamin, M., 1998. *The Garden of Ediacara: Discovering the First Complex Life*. Columbia University Press. 295pp. An idiosyncratic view of how complex life first evolved on the shores of the supercontinent Rodinia.

Oreskes, Naomi, 1999. *The Rejection of Continental Drift – Theory and Method in American Earth Science*. Oxford University Press. 420pp.

Insightful analysis of the true sociocultural reasons why US scientists found Wegener so hard to swallow.

Oreskes, Naomi, 2003. *Plate tectonics: An Insider's History of the Modern Theory of the Earth.* Westview Press. 424pp. Accounts in their own words by many of the surviving major players in the plate-tectonic revolution.

Ramaswamy, Sumathi, 2004. *The Lost Land of Lemuria: Fabulous Geographies, Catastrophic Histories.* University of California Press. 334pp. On the interface between Lemuria, Tamil world history and culture. Superb scholarly study by the noted Tamil cultural historian.

Rogers, John J. W., & Santosh, M., 2004. *Continents and Supercontinents.* Oxford University Press, New York. 289pp. Graduate/postgraduate level academic textbook for Earth scientists on the Supercontinent Cycle.

Vrielynck, Bruno, & Bouysse, Philippe, 2003. *The Changing Face of the Earth: The Break-up of Pangaea and Continental Drift over the Past 250 Million Years in Ten Steps.* Commission for the Geological Map of the World/UNESCO publishing. Available from CMGW, Paris http://ccgm.free.fr/ 33pp + CD ROM.

Winchester, Simon, 2005. *A Crack in the Edge of the World – The Great American Earthquake of 1906.* Viking Penguin. 412pp. Immensely readable account of the whys, wherefores and social context of the San Andreas Fault and San Francisco's greatest calamity.

Websites

www.thefutureiswild.com for information on the TV documentary and how to get hold of the CD set produced by Paramount Home Entertainment.

INDEX